This book provides in a single text detailed coverage of the variety of models that are currently being used in the empirical analysis of financial markets. Covering bond, equity and foreign exchange markets, it is aimed at scholars and practitioners wishing to acquire an understanding of the latest research techniques and findings in the field, and also at graduate students wishing to research in financial markets.

The book is divided into two main sections, covering univariate models, and econometric and multivariate techniques respectively. In the former, the areas covered include linear and non-linear stochastic models, random walks, unit root tests, GARCH models, deterministic chaos, trend reversion, and bubbles. In the latter, regression models, time-varying parameter models, the Kalman filter, vector autoregressions, present value models, and cointegration are discussed.

The econometric modelling of financial time series

The econometric modelling of financial time series

TERENCE C. MILLS
Professor of Applied Economics
Department of Economics
University of Hull

CAMBRIDGE
UNIVERSITY PRESS

Published by the Press Syndicate of the University of Cambridge
The Pitt Building, Trumpington Street, Cambridge CB2 1RP
40 West 20th Street, New York, NY 10011–4211, USA
10 Stamford Road, Oakleigh, Victoria 3166, Australia

First published 1993

Printed in Great Britain at the University Press, Cambridge

A catalogue record for this book is available from the British Library

Library of Congress cataloguing in publication data
Mills. Terence C.
The econometric modelling of financial time series / Terence C. Mills.
p. cm.
Includes bibliographical references and index.
ISBN 0-521-41048-7
1. Finance–Econometric models. 2. Time-series analysis.
3. Stochastic processes. I. Title.
HG174.M55 1993
332′.01′5195–dc20 92-37624 CIP

ISBN 0 521 41048 7 hardback

SE

Contents

1 Introduction

The aim of this book is to provide the researcher in financial markets with the techniques necessary to undertake the empirical analysis of financial time series. To accomplish this aim we introduce and develop both univariate modelling techniques and multivariate methods, including those regression techniques for time series that seem to be particularly relevant to the finance area.

Why do we concentrate exclusively on time series techniques when, for example, cross-sectional modelling plays an important role in empirical investigations of the Capital Asset Pricing Model (CAPM): see, as an early and influential example, Fama and MacBeth (1973)? Our answer is that, apart from the usual considerations of personal expertise and interest plus manuscript length constraints, it is because time series analysis, in both its theoretical and empirical aspects, has been for many years an integral part of the study of financial markets, with empirical research beginning with the papers by Working (1934), Cowles (1933, 1944) and Cowles and Jones (1937).

Working focused attention on a previously noted characteristic of commodity and stock prices: namely, that they resemble cumulations of purely random changes. Cowles investigated the ability of market analysts and financial services to predict future price changes, finding that there was little evidence that they could. Cowles and Jones reported evidence of positive correlation between successive price changes but, as Cowles (1960) was later to remark, this was probably due to their taking monthly averages of daily or weekly prices before computing changes: a 'spurious correlation' phenomenom analysed by Working (1960).

The predictability of price changes has since become a major theme of financial research but, surprisingly, little more was published until Kendall's (1953) study, in which he found that the weekly changes in a wide variety of financial prices could not be predicted from either past changes in the series themselves or from past changes in other price series. This seems to have been the first explicit reporting of this oft-quoted property of financial prices, although further impetus to research on price predictability was only provided by the publication of the papers by Roberts (1959) and Osborne (1959). The former presents a largely heuristic argument for why successive price changes should be independent, while the latter develops the proposition that it is not absolute price changes but the logarithmic price changes which are independent of each other: with the auxiliary assumption that the changes themselves are normally distributed, this implies that prices are generated as Brownian motion.

The stimulation provided by these papers was such that numerous articles appeared over the next few years investigating the hypothesis that price changes (or logarithmic price changes) are independent: an hypothesis that became to be termed the random walk model, in recognition of the similarity of the evolution of a price series to the random stagger of a drunk. Indeed, the term 'random walk' is believed to have first been used in an exchange of correspondence appearing in *Nature* in 1905 (see Pearson and Rayleigh, 1905), which was concerned about the optimal search strategy for finding a drunk who had been left in the middle of a field. The solution is to start exactly where the drunk had been placed, as that point is an unbiased estimate of the drunk's future position since he will presumably stagger along in an unpredictable and random fashion.

The most natural way to state formally the random walk model is as

$$P_t = P_{t-1} + a_t, \tag{1.1}$$

where P_t is the price observed at the beginning of time t and a_t is an error term which has zero mean and whose values are independent of each other. The price change, $\nabla P_t = P_t - P_{t-1}$, is thus simply a_t and is hence independent of past price changes. Note that, by successive backward substitution in (1.1), we can write the current price as the cumulation of all past errors, i.e.

$$P_t = \sum_{i=1}^{t} a_i,$$

so that the random walk model implies that prices are indeed generated by Working's 'cumulation of purely random changes'. Osborne's model of Brownian motion implies that equation (1.1) holds for the logarithms of P_t and, further, that a_t is drawn from a zero mean normal distribution having constant variance.

Most of the early papers in this area are contained in the collection of Cootner (1964), while Granger and Morgenstern (1970) provide a detailed development and empirical examination of the random walk model and various of its refinements. Amazingly, much of this work had been anticipated by the French mathematician Louis Bachelier (1900, English translation in Cootner, 1964) in a remarkable Ph.D. dissertation in which he developed an elaborate mathematical theory of speculative prices, which he then tested on the pricing of French government bonds, finding that such prices were consistent with the random walk model. What made the thesis even more remarkable was that it also developed many of the mathematical properties of Brownian motion which had been thought to have first been derived some years later in the physical sciences, particularly by Einstein! Yet, as Mandelbrot (1989) remarks, Bachelier had great difficulty in even getting himself a university appointment, let alone getting his theories disseminated throughout the academic community!

It should be emphasised that the random walk model is only an hypothesis about how financial prices move. One way in which it can be tested is by examining the autocorrelation properties of price changes: see, for example, Fama (1965). A more general perspective is to view (1.1) as a particular model within the class of autoregressive-integrated-moving average (ARIMA) models popularised by Box and Jenkins (1976). Chapter 2 thus develops the theory of such models within the general context of (univariate) linear stochastic processes.

We should avoid giving the impression that the only financial time series of interest are stock prices. There are financial markets other than those for stocks, most notably for bonds and foreign currency, but there also exist the various futures and commodity markets, all of which provide interesting and important series to analyse. For certain of these, it is by no means implausible that models other than the random walk may be appropriate or, indeed, models from a class other than the ARIMA. Chapter 3 discusses various topics in the general analysis of linear stochastic models, most notably that of determining the order of integration of a series and, associated with this, the appropriate way of modelling trends and structural breaks, but it also considers related methods of decomposing an observed series into two or more unobserved components and of determining the extent of the 'memory' of a series, by which is meant the behaviour of the series at low frequencies or, equivalently, in the very long run. A variety of examples taken from the financial literature are provided throughout the chapter.

During the 1960s much research was also carried out on the theoretical foundations of financial markets, leading to the development of the theory of efficient capital markets. As LeRoy (1989) discusses, this led to some serious questions being raised about the random walk hypothesis as a *theoretical* model of financial markets, the resolution of which required situating the hypothesis within a framework of economic equilibrium. Unfortunately, the assumption in (1.1) that price changes are independent was found to be too restrictive to be generated within a reasonably broad class of optimising models. A model that is appropriate, however, can be derived for stock prices in the following way (similar models can be derived for other sorts of financial prices, although the justification is sometimes different: see LeRoy, 1982). The return on a stock from t to $t+1$ is defined as the sum of the dividend yield and the capital gain, i.e., as

$$r_{t+1}=\frac{P_{t+1}+D_t-P_t}{P_t}, \tag{1.2}$$

where D_t is the dividend paid during period t. Let us suppose that the expected return is constant, $E_t(r_{t+1})=r$: r_t is then said to be a *fair game*. Taking expectations at time t of both sides of (1.2) and rearranging yields

$$P_t=(1+r)^{-1}E_t(P_{t+1}+D_t), \tag{1.3}$$

which says that the stock price at the beginning of period t equals the sum of the expected future price and dividend, discounted back at a rate r. Now assume that there is a mutual fund that holds the stock in question and that it reinvests dividends in future share purchases. Suppose that it holds h_t shares at the beginning of period t, so that the value of the fund is $x_t=h_tP_t$. The assumption that the fund ploughs back its dividend income implies that h_{t+1} satisfies

$$h_{t+1}P_{t+1}=h_t(P_{t+1}+D_t).$$

Thus

$$E_t(x_{t+1})=E_t(h_{t+1}P_{t+1})=h_tE_t(P_{t+1}+D_t)=(1+r)h_tP_t=(1+r)x_t,$$

i.e., that x_t is a *martingale* (if, as is common, $r>0$, we have that $E_t(x_{t+1})\geq x_t$, so that x_t is a *submartingale*: LeRoy (1989, pp. 1593–4), however, offers an example in which r could be negative, in which case x_t will be a *supermartingale*). LeRoy (1989) emphasises that price itself, without dividends added in, is not generally a martingale: from (1.3) we have

$$r=E_t(D_t)/P_t+E_t(P_{t+1})/P_t-1,$$

so that only if the expected dividend–price ratio is constant, say $E_t(D_t)/P_t=d$, can we write P_t as the submartingale (assuming $r>d$)

$$E_t(P_{t+1})=(1+r-d)/P_t.$$

The assumption that a stochastic process, y_t say, follows a random walk is more restrictive than the requirement that y_t follows a martingale. The martingale rules out any dependence of the conditional expectation of ∇y_{t+1} on the information available at t, whereas the random walk rules out not only this but also dependence involving the higher conditional moments of ∇y_{t+1}. The importance of this distinction is thus evident: financial series are known to go through protracted quiet periods and also protracted periods of turbulence. This type of behaviour could be modelled by a process in which successive conditional variances of ∇y_{t+1} (but *not* successive levels) are positively autocorrelated: such a specification is consistent with a martingale, but not with the more restrictive random walk.

Martingale processes are discussed in chapter 4, and lead naturally on to non-linear stochastic processes that are capable of modelling higher conditional moments, such as the autoregressive conditionally heteroskedastic (ARCH) model introduced by Engle (1982) and the bilinear process analysed by Granger and Andersen (1978). Also discussed in this chapter are non-linear Markov processes, chaotic and asymmetric models, and the various tests of non-linearity that have recently been developed. The techniques are illustrated using exchange rates and stock price series.

Samuelson (1965, 1973) and Mandelbrot (1966) analysed the implications of equation (1.3), that the stock price at the beginning of period t equals the discounted sum of next periods expected future price and dividend, to show that this stock price equals the expected discounted, or present, value of *all* future dividends

$$P_t = \sum_{i=0}^{\infty} (1+r)^{i+1} E_t(D_{t+i}), \tag{1.4}$$

which is obtained by recursively solving (1.3) forwards and assuming that $(1+r)^n E_t(P_{t+n})$ converges to zero as $n \rightarrow \infty$. Present value models of the type (1.4) are analysed extensively in chapter 7, with the theme of whether stock markets are excessively volatile, perhaps containing speculative 'bubbles', being used throughout the discussion and in a succession of examples, although the testing of the expectations hypothesis of the term structure of interest rates is also used as an example of the general present value framework. The chapter also considers a multivariate extension of ARCH models.

The analysis of such models requires the use of multivariate techniques of time series analysis, including regression methods, and these are developed prior to this in chapters 5 and 6. The former chapter concentrates on analysing the relationships between a set of *stationary* or, more precisely, *non-integrated*, financial time series and considers such topics as general dynamic regression models, ARCH-in-mean regression models, misspecification testing, multivariate regression models, vector autoregressions, Granger-causality and variance decompositions. These topics are illustrated with a variety of examples drawn from the financial literature: using forward exchange rates as optimal predictors of future spot rates, modelling the volatility of stock returns and the risk premium in the foreign exchange market, testing the CAPM, and investigating the transmission of shocks across world bond markets.

Chapter 6 concentrates on the modelling of *integrated* financial time series, beginning with a discussion of the spurious regression problem, then introducing cointegrated processes and demonstrating how to test for cointegration, and finally moving on to consider alternative representations of cointegrated systems and how they can be estimated. The techniques introduced in this chapter are illustrated with examples drawn from the interaction of world bond markets and the relationship between the stock and gilt markets in the UK.

Finally, there is a short chapter 8 setting out our views as to where we see future developments in the subject appearing and how the material covered in the book is becoming increasingly important within the wider financial and investment community.

Having emphasised earlier in this chapter that the book is exclusively about modelling financial time series, we should state at this juncture what the book is not about. It is certainly not a text on financial market theory: such theory is only discussed when it is necessary as a motivation for a particular technique or example. A number of texts on the theory of finance have recently been published (see, for example, Ingersoll, 1987, and Huang and Litzenberger, 1988), and the reader is referred to such texts for the requisite financial theory. Neither is it a textbook on econometrics. We assume that the reader already has a working knowledge of probability, statistics and econometric theory. Nevertheless, it is also non-rigorous, being at a level similar to my *Time Series Techniques for Economists* (1990) in which references to the formal treatment of the theory of time series are provided.

When the data used in the examples throughout the book have already been published, references are given. Previously unpublished data are listed in the data appendix. All standard regression computations were carried out using either *MICROTSP*, Version 7.0, or *MICROFIT*, Version 3.0. 'Non-standard' computations were made using algorithms written by the author in *GAUSS*, Version 2.0.

2 Univariate linear stochastic models: basic concepts

Chapter 1 has emphasised the standard representation of a financial time series as that of a (univariate) linear stochastic process, specifically as being a member of the class of ARIMA models popularised by Box and Jenkins (1976). This chapter thus provides the basic theory of such models within the general framework of the analysis of linear stochastic processes. As already stated in chapter 1, our treatment is purposely non-rigourous: for detailed theoretical treatments, but which do not, however, focus on the analysis of financial series, see, for example, Anderson (1971), Fuller (1976) and Brockwell and Davis (1991).

2.1 Stochastic processes, ergodicity and stationarity

2.1.1 Stochastic processes, realisations and ergodicity

When we wish to analyse a financial time series using formal statistical methods, it is useful to regard the observed series, $(x_1, x_2, \ldots, x_T)$, as a particular *realisation* of a stochastic process. This realisation is often denoted $\{x_t\}_1^T$ while, in general, the stochastic process will be the family of random variables $\{X_t\}_{-\infty}^{\infty}$ defined on an appropriate probability space. For our purposes it will usually be sufficient to restrict the index set $\mathcal{T} = (-\infty, \infty)$ of the parent stochastic process to be the same as that of the realisation, i.e., $\mathcal{T} = (1, T)$, and also to use x_t to denote both the stochastic process and the realisation when there is no possibility of confusion.

With these conventions, the stochastic process can be described by a T-dimensional probability distribution, so that the relationship between a realisation and a stochastic process is analogous to that between the sample and population in classical statistics. Specifying the complete form of the probability distribution will generally be too ambitious a task and we usually content ourselves with concentrating attention on the first and second moments: the T means

$$E(x_1),\ E(x_2), \ldots, E(x_T),$$

T variances

$$V(x_1),\ V(x_2), \ldots, V(x_T),$$

and $T(T-1)/2$ covariances

$$Cov(x_i, x_j),\ i<j.$$

If we could assume joint normality of the distribution, this set of expectations would then completely characterise the properties of the stochastic process. As we shall see, however, such an assumption is unlikely to be appropriate for many financial series. If normality cannot be assumed, but the process is taken to be *linear*, in the sense that the current value of the process is generated by a linear combination of previous values of the process itself and current and past values of any other related processes, then again this set of expectations would capture its major properties. In either case, however, it will be impossible to infer all the values of the first and second moments from just one realisation of the process, since there are only T observations but $T+T(T+1)/2$ unknown parameters. Hence further simplifying assumptions must be made to reduce the number of unknown parameters to more manageable proportions.

We should emphasise that the procedure of using a single realisation to infer the unknown parameters of a joint probability distribution is only valid if the process is *ergodic*, which roughly means that the sample moments for finite stretches of the realisation approach their population counterparts as the length of the realisation becomes infinite. For more on ergodicity see, for example, Granger and Newbold (1977, chapter 1) and Nerlove, Grether and Carvalho (1979, chapter 2), and since it is impossible to test for ergodicity using just (part of) a single realisation, it will be assumed from now on that all time series have this property.

2.1.2 Stationarity

One important simplifying assumption is that of *stationarity*, which requires the process to be in a particular state of 'statistical equilibrium' (Box and Jenkins, 1976, p. 26). A stochastic process is said to be *strictly stationary* if its properties are unaffected by a change of time origin: in other words, the joint probability distribution at *any* set of times $t_1, t_2, \ldots, t_m$ must be the same as the joint probability distribution at times $t_1+k, t_2+k, \ldots, t_m+k$, where k is an arbitrary shift along the time axis. For $m=1$, this implies that the marginal probability distributions do not depend on time, which in turn implies that, so long as $E|x_t|^2<\infty$, both the mean and variance of x_t must be constant, i.e.

$$E(x_1)=E(x_2)=\ldots=E(x_T)=E(x_t)=\mu,$$

and

$$V(x_1)=V(x_2)=\ldots=V(x_T)=V(x_t)=\sigma_x^2.$$

If $m=2$, strict stationarity implies that all bivariate distributions do not depend on t: thus all covariances are functions only of the lag k, and not of time t, i.e., for all k

$$Cov(x_1,x_{1+k})=Cov(x_2,x_{2+k})=\ldots=Cov(x_{T-k},x_T)=Cov(x_t,x_{t-k}).$$

In summary, then, the strict stationarity assumption implies that the mean and variance of the process are constant and that the *autocovariances*

$$\gamma_k=Cov(x_t,x_{t-k})=E[(x_t-\mu)(x_{t-k}-\mu)],$$

and *autocorrelations*,

$$\rho_k=\frac{Cov(x_t,x_{t-k})}{[V(x_t)\cdot V(x_{t-k})]^{\frac{1}{2}}}=\frac{\gamma_k}{\gamma_0},$$

depend only on the lag (or time difference) k. Since these conditions apply only to the first- and second-order moments of the process, this is known as *second-order* or *weak stationarity* (and sometimes *covariance stationarity* or *stationarity in the wide sense*). While strict stationarity (with finite second moments) thus implies weak stationarity, the converse does not hold, for it is possible for a process to be weakly stationary but *not* strictly stationary. If, however, joint normality can be assumed, so that the distribution is entirely characterised by these first two moments, weak stationarity does indeed imply strict stationarity.

The autocorrelations considered as a function of k are referred to as the *autocorrelation function* (ACF). Note that since

$$\gamma_k=Cov(x_t,x_{t-k})=Cov(x_{t-k},x_t)=Cov(x_t,x_{t+k})=\gamma_k,$$

it follows that $\rho_k=\rho_{-k}$ and so only the positive half of the ACF is usually given. The ACF plays a major role in modelling dependencies among observations since it characterises, together with the process mean $\mu=E(x_t)$ and variance $\sigma_x^2=\gamma_0=V(x_t)$, the stationary stochastic process describing the evolution of x_t. It therefore indicates, by measuring the extent to which one value of the process is correlated with previous values, the length and strength of the 'memory' of the process.

2.2 Stochastic difference equations

A fundamental theorem in time series analysis, known as *Wold's decomposition* (Wold, 1938), states that every weakly stationary, purely non-deterministic, stochastic process $(x_t - \mu)$ can be written as a linear combination (or *linear filter*) of a sequence of uncorrelated random variables. By purely non-deterministic we mean that any linearly deterministic component has been subtracted from x_t. Such a component is one that can be perfectly predicted from past values of itself and examples commonly found are a (constant) mean, as is implied by writing the process as $(x_t - \mu)$, periodic sequences, and polynomial or exponential sequences in t. A formal discussion of this theorem, well beyond the scope of this book, may be found in, for example, Brockwell and Davis (1991, chapter 5.7), but Wold's decomposition underlies all the theoretical models of time series that are subsequently to be introduced.

This linear filter representation is given by

$$\begin{aligned} x_t - \mu &= a_t + \psi_1 a_{t-1} + \psi_2 a_{t-2} + \ldots \\ &= \sum_{j=0}^{\infty} \psi_j a_{t-j}, \quad \psi_0 = 1 . \end{aligned} \tag{2.1}$$

The $\{a_t: t = 0, \pm 1, \pm 2, \ldots\}$ are a sequence of uncorrelated random variables, often known as *innovations*, drawn from a fixed distribution with

$$E(a_t) = 0,\ V(a_t) = E(a_t^2) = \sigma^2 < \infty ,$$

and

$$Cov(a_t, a_{t-k}) = E(a_t a_{t-k}) = 0, \text{ for all } k \neq 0.$$

We will refer to such a sequence as a *white-noise* process, often denoting it as $a_t \sim WN(0, \sigma^2)$. The coefficients (possibly infinite in number) in the linear filter are known as *ψ-weights*.

We can easily show that the model (2.1) leads to autocorrelation in x_t. From this equation it follows that

$$\begin{aligned} E(x_t) &= \mu \\ \gamma_0 &= V(x_t) = E(x_t - \mu)^2 \\ &= E(a_t + \psi_1 a_{t-1} + \psi_2 a_{t-2} + \ldots)^2 \\ &= E(a_t^2) + \psi_1^2 E(a_{t-1}^2) + \psi_2^2 E(a_{t-2}^2) + \ldots \\ &= \sigma^2 + \psi_1^2 \sigma^2 + \psi_2^2 \sigma^2 + \ldots \\ &= \sigma^2 \sum_{j=0}^{\infty} \psi_j^2 , \end{aligned}$$

by using the result that $E(a_{t-i} a_{t-j}) = 0$ for $i \neq j$. Now

$$\begin{aligned} \gamma_k &= E(x_t - \mu)(x_{t-k} - \mu) \\ &= E(a_t + \psi_1 a_{t-1} + \ldots + \psi_k a_{t-k} + \psi_{k+1} a_{t-k-1} + \ldots) \\ &\quad \times (a_{t-k} + \psi_1 a_{t-k-1} + \ldots) \end{aligned}$$

$$= \sigma^2(1 \cdot \psi_k + \psi_1\psi_{k+1} + \psi_2\psi_{k+2} + \ldots)$$

$$= \sigma^2 \sum_{j=0}^{\infty} \psi_j\psi_{j+k},$$

and this implies

$$E_k = \frac{\sum_{j=0}^{\infty} \psi_j\psi_{j+k}}{\sum_{j=0}^{\infty} \psi_j^2}.$$

If the number of ψ-weights in (2.1) is infinite, we have to assume that the weights converge absolutely ($\sum |\psi_j| < \infty$). This condition can be shown to be equivalent to assuming that x_t is stationary, and guarantees that all moments exist and are independent of time, in particular that the variance of x_t, γ_0, is finite.

2.3 ARMA processes

2.3.1 Autoregressive processes

Although equation (2.1) may appear complicated, many realistic models result from particular choices of the ψ-weights. Taking $\mu = 0$ without loss of generality, choosing $\psi_j = \phi^j$ allows (2.1) to be written

$$\begin{aligned} x_t &= a_t + \phi a_{t-1} + \phi^2 a_{t-2} + \ldots \\ &= a_t + \phi(a_{t-1} + \phi a_{t-2} + \ldots) \\ &= \phi x_{t-1} + a_t, \end{aligned}$$

or

$$x_t - \phi x_{t-1} = a_t. \tag{2.2}$$

This is known as a *first-order autoregressive* process, often given the acronym AR(1). The *backshift* (or *lag*) *operator* B is now introduced for notational convenience. This shifts time one step back, so that

$$Bx_t \equiv x_{t-1},$$

and, in general

$$B^m x_t \equiv x_{t-m},$$

noting that $B^m \mu \equiv \mu$. The lag operator allows (possibly infinite) distributed lags to be written in a very concise way. For example, by using this notation the AR(1) model can be written as

$$(1 - \phi B)x_t = a_t,$$

so that

$$x_t = (1-\phi B)^{-1}a_t = (1+\phi B+\phi^2B^2+\ldots)a_t$$
$$= a_t+\phi a_{t-1}+\phi^2 a_{t-2}+\ldots \qquad (2.3)$$

This linear filter representation will converge as long as $|\phi|<1$, which is therefore the stationarity condition.

We can now deduce the ACF of an AR(1) process. Multiplying both sides of (2.2) by x_{t-k}, $k\geq 0$, and taking expectations yields

$$\gamma_k - \phi\gamma_{k-1} = E(a_t x_{t-k}). \qquad (2.4)$$

From (2.3), $a_t x_{t-k} = \sum_{i=0}^{\infty}\phi^i a_t a_{t-k-i}$. As $\{a_t\}$ is white noise, any term in $a_t a_{t-k-i}$ has zero expectation if $k+i>0$. Thus (2.4) simplifies to

$$\gamma_k = \phi\gamma_{k-1}, \text{ for all } k>0,$$

and, consequently, $\gamma_k = \phi^k\gamma_0$. An AR(1) process therefore has an ACF given by $\rho_k = \phi^k$. Thus, if $\phi>0$, the ACF decays exponentially to zero, while if $\phi<0$, the ACF decays in an oscillatory pattern, both decays being slow if ϕ is close to the non-stationary boundaries of $+1$ and -1.

The ACFs for two AR(1) processes with (a) $\phi=0.5$, and (b) $\phi=-0.5$, are shown in figure 2.1, along with generated data from the processes with a_t assumed to be normally and independently distributed with $\sigma^2=25$, denoted $a_t \sim NID(0,25)$, and with starting value $x_0=0$. With $\phi>0$ the adjacent values are positively correlated and the generated series has a tendency to exhibit 'low-frequency' trends. With $\phi<0$, however, adjacent values have a negative correlation and the generated series displays violent, rapid oscillations.

2.3.2 Moving average processes

Now consider the model obtained by choosing $\psi_1 = -\theta$ and $\psi_j=0, j\geq 2$, in (2.1):

$$x_t = a_t - \theta a_{t-1},$$

or (2.5)

$$x_t = (1-\theta B)a_t.$$

This is known as the *first-order moving average* [MA(1)] process and it follows immediately that

$$\gamma_0 = \sigma^2(1+\theta^2),\ \gamma_1 = -\sigma^2\theta,\ \gamma_k = 0 \text{ for } k>1,$$

and hence its ACF is described by

$$\rho_1 = \frac{-\theta}{1+\theta^2},\ \rho_k = 0, \text{ for } k>1.$$

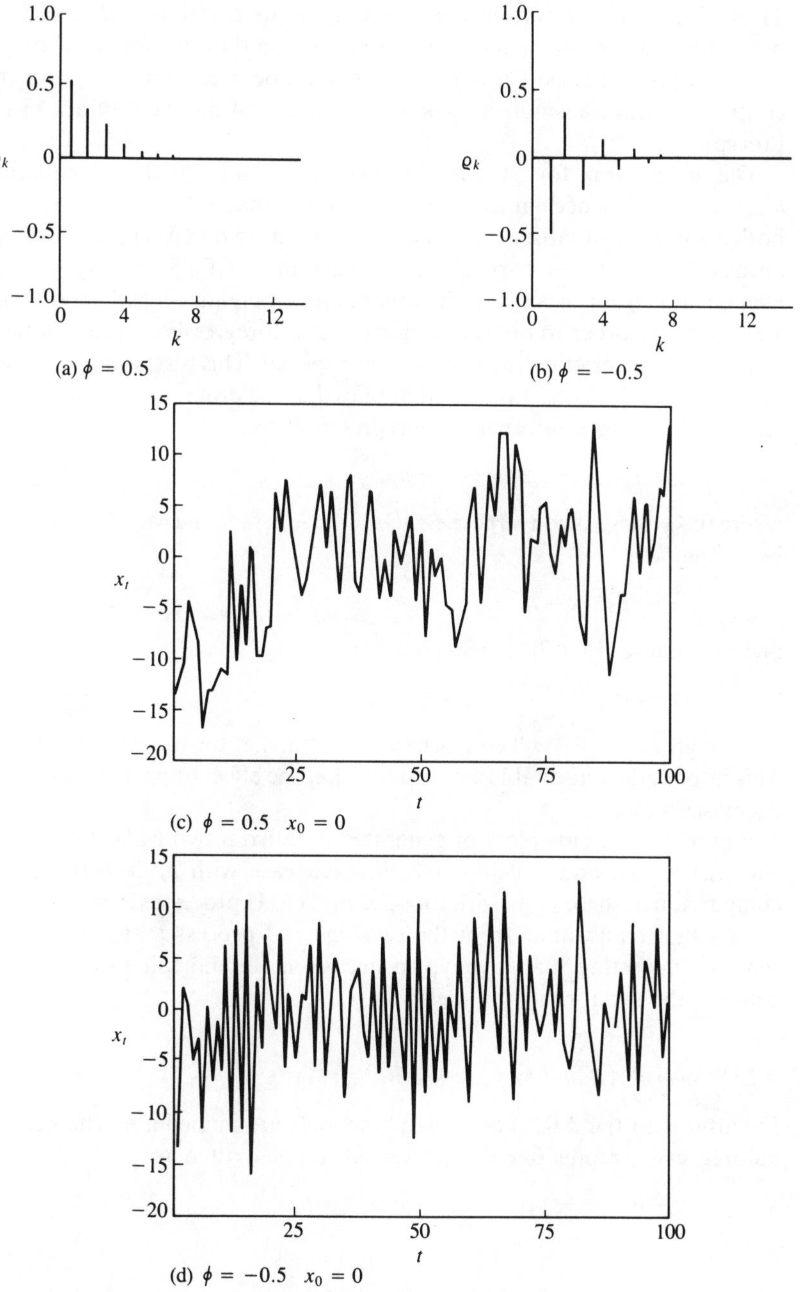

Figure 2.1 ACFs and simulations of AR(1) processes

Thus, although observations one period apart are correlated, observations more than one period apart are uncorrelated, so that the 'memory' of the process is just one period: this 'jump' to zero autocorrelation at $k=2$ may be contrasted with the smooth, exponential, decay of the ACF of an AR(1) process.

The expression for ρ_1 can be written as the quadratic equation $\theta^2\rho_1+\theta+\rho_1=0$. Since θ must be real, it follows that $-\frac{1}{2}<\rho_1<\frac{1}{2}$. However, both θ and $1/\theta$ will satisfy this equation and thus two MA(1) processes can always be found that correspond to the same ACF. Since any moving average model consists of a finite number of ψ-weights, all MA models are stationary. In order to obtain a converging autoregressive representation, however, the restriction $|\theta|<1$ must be imposed. This restriction is known as the *invertibility* condition and it implies that the process can be written in terms of an infinite autoregressive representation

$$x_t=\pi_1x_{t-1}+\pi_2x_{t-2}+\ldots+a_t,$$

where the *π-weights* converge, i.e., $\sum|\pi_j|<\infty$. In fact, the MA(1) model can be written as

$$(1-\theta B)^{-1}x_t=a_t,$$

and expanding $(1-\theta B)^{-1}$ yields

$$(1+\theta B+\theta^2B^2+\ldots)x_t=a_t.$$

The weights $\pi_j=-\theta^j$ will converge if $|\theta|<1$, i.e., if the model is invertible. This implies the reasonable assumption that the effect of past observations decreases with age.

Figure 2.2 presents plots of generated data from two MA(1) processes with (a) $\theta=0.8$ and (b) $\theta=-0.8$, in each case with $a_t \sim NID(0,25)$. On comparison of these plots with those of the AR(1) processes in figure 2.1, it is seen that realisations from the two types of processes are often quite similar, suggesting that it may, on occasions, be difficult to distinguish between the two forms of models.

2.3.3 General AR and MA processes

Extensions to the AR(1) and MA(1) models are immediate. The general autoregressive model of order p [AR(p)] can be written as

$$x_t-\phi_1x_{t-1}-\phi_2x_{t-2}-\ldots-\phi_px_{t-p}=a_t,$$

or

$$(1-\phi_1B-\phi_2B^2-\ldots-\phi_pB^p)x_t=a_t,$$

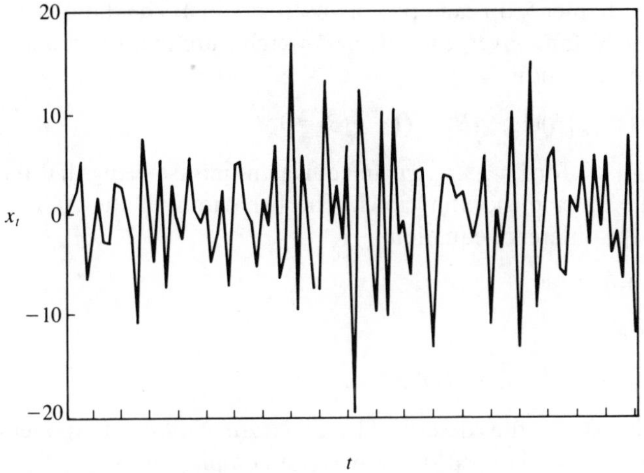

(a) $\theta = 0.8$

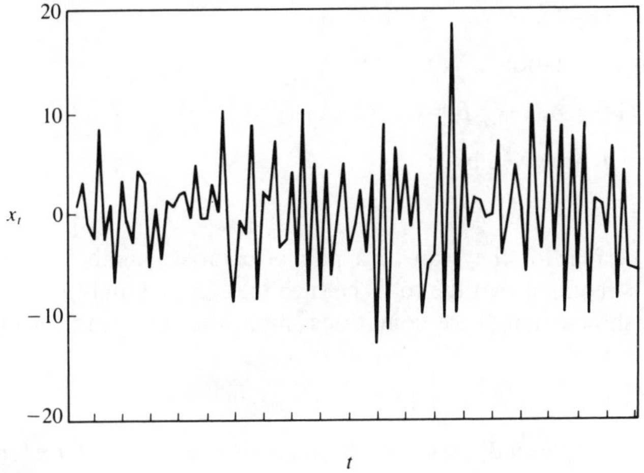

(b) $\theta = -0.8$

Figure 2.2 Simulations of MA(1) processes

i.e.

$$\phi(B)x_t = a_t.$$

The linear filter representation $x_t = \psi(B)a_t$ can be obtained by equating coefficients in

$$\phi(B)\psi(B) = 1$$

(see Mills, 1990, chapter 5, for examples of how to do this). The stationarity conditions required for convergence of the ψ-weights are that the roots of the characteristic equation

$$\phi(B) = (1 - g_1 B)(1 - g_2 B) \ldots (1 - g_p B) = 0$$

are such that $|g_i| < 1$ for $i = 1,2,\ldots,p$, an equivalent phrase being that the roots g_i^{-1} all lie outside the unit circle. The behaviour of the ACF is determined by the difference equation

$$\phi(B)\rho_k = 0\,, \; k > 0\,, \tag{2.6}$$

which has the solution

$$\rho_k = A_1 g_1^k + A_2 g_2^k + \ldots + A_p g_p^k\,.$$

Since $|g_i| < 1$, the ACF is thus described by a mixture of damped exponentials (for real roots) and damped sine waves (for complex roots).

As an example, consider the AR(2) process

$$(1 - \phi_1 B - \phi_2 B^2) x_t = a_t\,,$$

with characteristic equation

$$\phi(B) = (1 - g_1 B)(1 - g_2 B) = 0\,.$$

The roots g_1 and g_2 are given by

$$g_1, g_2 = (\phi_1 \pm (\phi_1^2 + 4\phi_2)^{\frac{1}{2}})/2$$

and can both be real, or they can be a pair of complex numbers. For stationarity, it is required that the roots be such that $|g_1| < 1$ and $|g_2| < 1$, and it can be shown that these conditions imply the following set of restrictions on ϕ_1 and ϕ_2

$$\phi_1 + \phi_2 < 1, \qquad -\phi_1 + \phi_2 < 1, \qquad -1 < \phi_2 < 1\,.$$

The roots will be complex if $\phi_1^2 + 4\phi_2 < 0$, although a necessary condition for complex roots is simply that $\phi_2 < 0$.

The behaviour of the ACF of an AR(2) process for four combinations of (ϕ_1, ϕ_2) is shown in figure 2.3. If g_1 and g_2 are real (cases (a) and (b)), the ACF is a mixture of two damped exponentials. Depending on their sign, the autocorrelations can also damp out in an oscillatory manner. If the roots are complex (cases (c) and (d)), the ACF follows a damped sine wave. Figure 2.4 shows plots of generated time series from these four AR(2) processes, in each case with $a_t \sim NID(0,25)$. Depending on the signs of the real roots, the series may be either smooth or jagged, while complex roots tend to induce 'pseudo-periodic' behaviour.

Since all AR processes have ACFs that 'damp out', it is sometimes

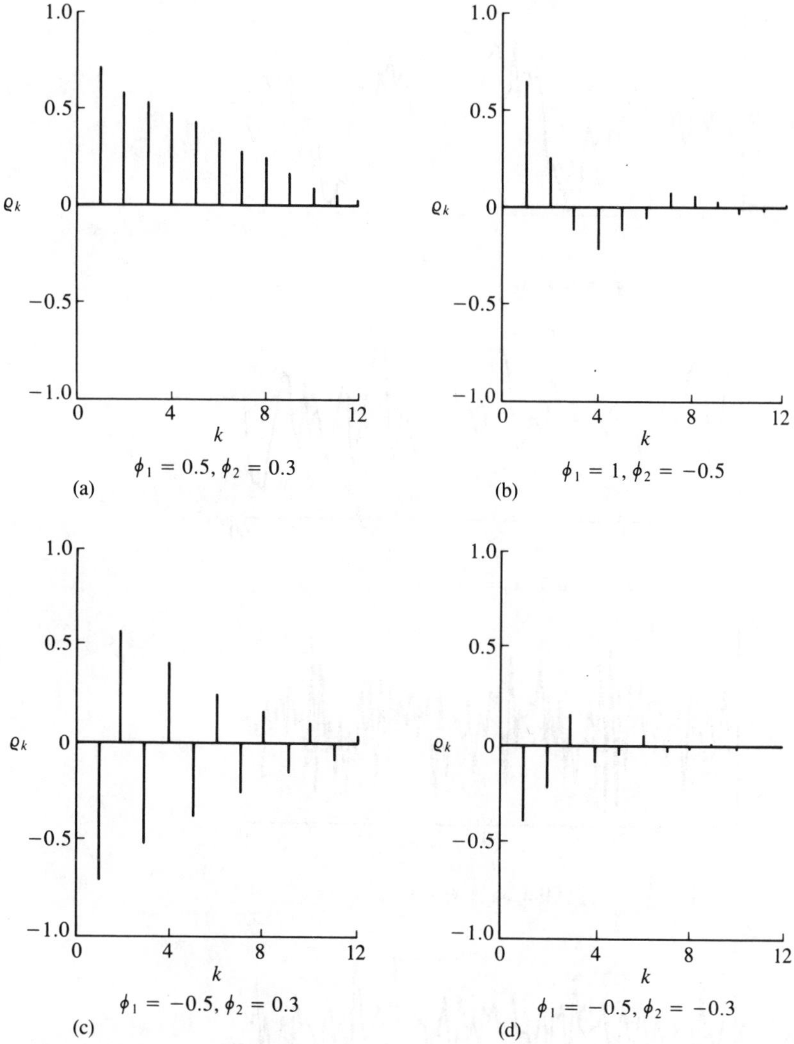

Figure 2.3 ACFs of various AR(2) processes

difficult to distinguish between processes of different orders. To aid with such discrimination, we may use the *partial autocorrelation function* (PACF). In general, the correlation between two random variables is often due to both variables being correlated with a third. In the present context, a large portion of the correlation between x_t and x_{t-k} may be due to the correlation this pair have with the intervening lags $x_{t-1}, x_{t-2}, \ldots, x_{t-k+1}$. To adjust for this correlation, the *partial autocorrelations* may be calculated.

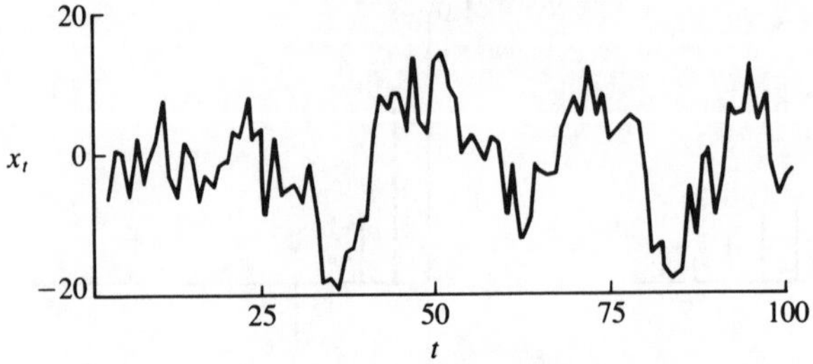

(a) $\phi_1 = 0.5 \quad \phi_2 = 0.3 \quad x_0 = x_1 = 0$

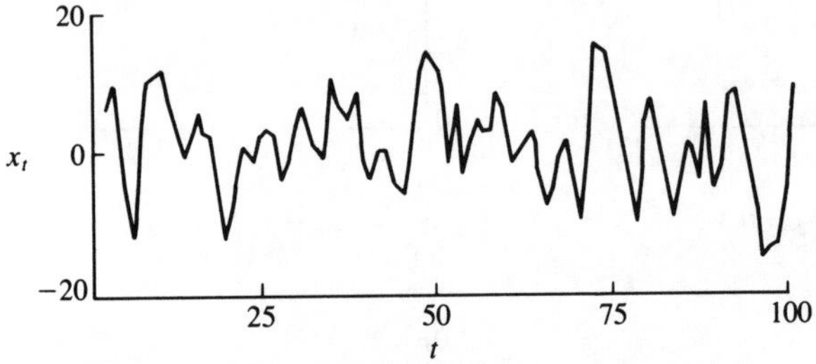

(b) $\phi_1 = 1 \quad \phi_2 = -0.5 \quad x_0 = x_1 = 0$

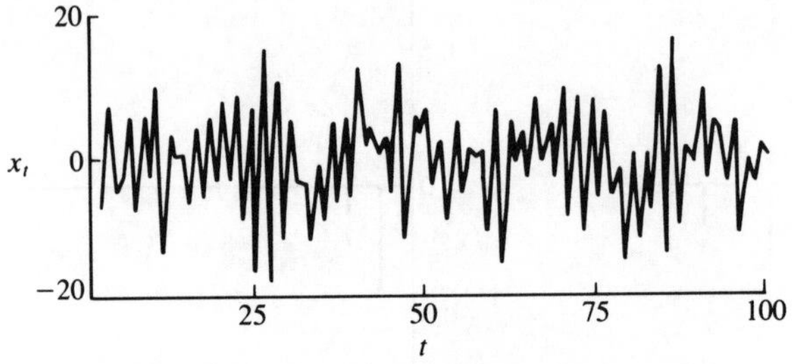

(c) $\phi_1 = -0.5 \quad \phi_2 = 0.3 \quad x_0 = x_1 = 0$

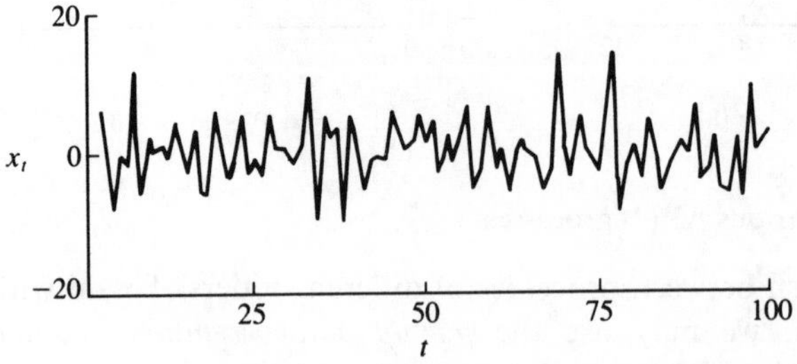

(d) $\phi_1 = -0.5 \quad \phi_2 = -0.3 \quad x_0 = x_1 = 0$

Figure 2.4 Simulations of various AR(2) processes

The kth partial autocorrelation is the coefficient ϕ_{kk} in the AR(k) process

$$x_t = \phi_{k1}x_{t-1} + \phi_{k2}x_{t-2} + \ldots + \phi_{kk}x_{t-k} + a_t, \tag{2.7}$$

and measures the additional correlation between x_t and x_{t-k} after adjustments have been made for the intervening lags.

In general, ϕ_{kk} can be obtained from the *Yule-Walker* equations that correspond to (2.7). These are given by the set of equations (2.6) with $p=k$ and $\phi_i=\phi_{ii}$, and solving for the last coefficient ϕ_{kk} using Cramer's Rule leads to

$$\phi_{kk} = \frac{\begin{vmatrix} 1 & \rho_1 & \cdots & \rho_{k-2} & \rho_1 \\ \rho_1 & 1 & \cdots & \rho_{k-3} & \rho_2 \\ \cdot & \cdot & \cdots & \cdot & \cdot \\ \cdot & \cdot & \cdots & \cdot & \cdot \\ \cdot & \cdot & \cdots & \cdot & \cdot \\ \rho_{k-1} & \rho_{k-2} & \cdots & \rho_1 & \rho_k \end{vmatrix}}{\begin{vmatrix} 1 & \rho_1 & \cdots & \rho_{k-2} & \rho_{k-1} \\ \rho_1 & 1 & \cdots & \rho_{k-3} & \rho_{k-2} \\ \cdot & \cdot & \cdots & \cdot & \cdot \\ \cdot & \cdot & \cdots & \cdot & \cdot \\ \cdot & \cdot & \cdots & \cdot & \cdot \\ \rho_{k-1} & \rho_{k-2} & \cdots & \rho_1 & 1 \end{vmatrix}}.$$

It follows from the definition of ϕ_{kk} that the PACFs of AR processes are of a particular form:

$$\text{AR}(1)\text{: } \phi_{11}=\rho_1=\phi, \qquad \phi_{kk}=0 \text{ for } k>1.$$
$$\text{AR}(2)\text{: } \phi_{11}=\rho_1,\ \phi_{22}=\frac{\rho_2-\rho_1^2}{1-\rho_1^2}, \qquad \phi_{kk}=0 \text{ for } k>2.$$
$$\text{AR}(p)\text{: } \phi_{11}\neq 0,\ \phi_{22}\neq 0, \ldots, \phi_{pp}\neq 0, \quad \phi_{kk}=0 \text{ for } k>p.$$

Thus the partial autocorrelations for lags larger than the order of the process are zero. Hence an AR(p) process is described by:

(i) an ACF that is infinite in extent and is a combination of damped exponentials and damped sine waves, and

(ii) a PACF that is zero for lags larger than p.

The general moving average model of order q [MA(q)] can be written as

$$x_t = a_t - \theta_1 a_{t-1} - \ldots - \theta_q a_{t-q},$$

or

$$x_t = (1-\theta_1 B - \ldots - \theta_q B^q)a_t = \theta(B)a_t.$$

The ACF can be shown to be

$$\rho_k = \frac{-\theta_k + \theta_1\theta_{k+1} + \ldots + \theta_{q-k}\theta_q}{1+\theta_1^2+\ldots+\theta_q^2}, \quad k=1,2,\ldots,q,$$

$$\rho_k = 0,\ k>q.$$

The ACF of an MA(q) process therefore cuts off after lag q: the memory of the process extends q periods, observations more than q periods apart being uncorrelated.

The weights in the AR(∞) representation $\pi(B)x_t = a_t$ are given by $\pi(B) = \theta^{-1}(B)$ and can be obtained by equating coefficients of B^j in $\pi(B)\theta(B) = 1$. For invertibility, the roots of

$$(1-\theta_1 B - \ldots - \theta_q B^q) = (1-h_1 B)\ldots(1-h_q B) = 0$$

must satisfy $|h_i| < 1$ for $i = 1,2,\ldots,q$.

Figure 2.5 presents generated series from two MA(2) processes, again using $a_t \sim NID(0,25)$. The series tend to be fairly jagged, similar to AR(2) processes with real roots of opposite signs, and, of course, such MA processes are unable to capture periodic-type behaviour.

The PACF of an MA(q) process can be shown to be infinite in extent (i.e., it tails off). Explicit expressions for the PACF of MA processes are complicated but, in general, are dominated by combinations of exponential decays (for the real roots in $\theta(B)$) and/or damped sine waves (for the complex roots). Their patterns are thus very similar to the ACFs of AR processes. Indeed, an important duality between AR and MA processes exists: while the ACF of an AR(p) process is infinite in extent, the PACF cuts off after lag p. The ACF of an MA(q) process, on the other hand, cuts off after lag q, while the PACF is infinite in extent.

2.3.4 Autoregressive-moving average models

We may also consider combinations of autoregressive and moving average models. For example, consider the natural combination of the AR(1) and MA(1) models, known as the *first-order autoregressive-moving average* model, or ARMA(1,1)

$$x_t - \phi x_{t-1} = a_t - \theta a_{t-1},$$

or (2.8)

$$(1-\phi B)x_t = (1-\theta B)a_t.$$

The ψ-weights in the MA(∞) representation are given by

$$\psi(B) = \frac{(1-\theta B)}{(1-\phi B)},$$

i.e.

$$x_t = \psi(B)a_t = \left(\sum_{i=0}^{\infty}\phi_i B^i\right)(1-\theta B)a_t = a_t + (\phi-\theta)\sum_{i=0}^{\infty}\phi^{i-1}a_{t-i}. \quad (2.9)$$

(a) $\theta_1 = -0.5, \theta_2 = 0.3$

(b) $\theta_1 = 0.5, \theta_2 = 0.3$

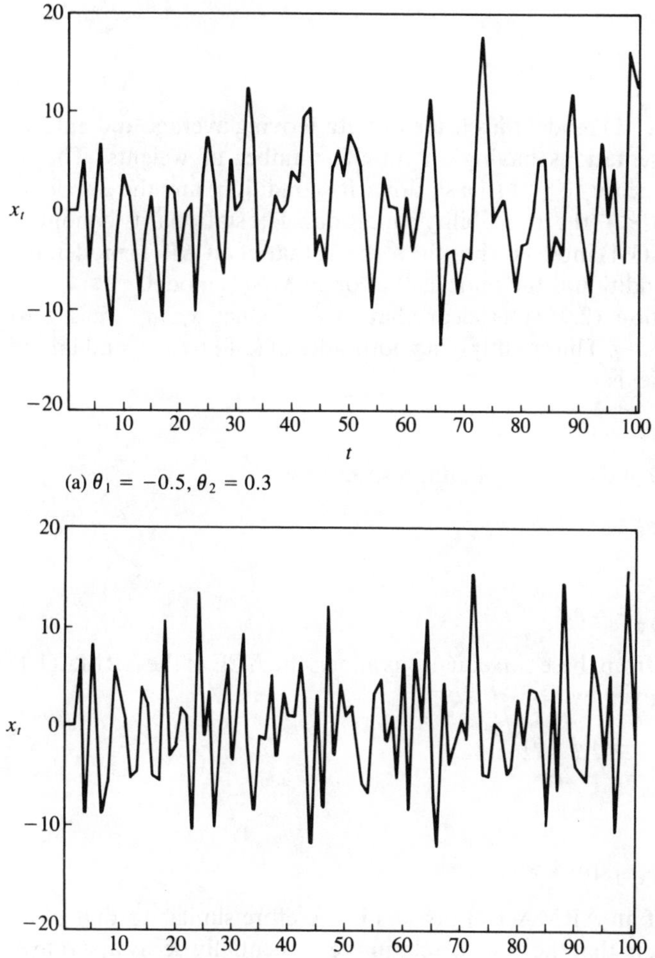

Figure 2.5 Simulations of MA(2) processes

Likewise, the π-weights in the AR(∞) representation are given by

$$\pi(B) = \frac{(1 - \phi B)}{(1 - \theta B)},$$

i.e.

$$\pi(B)x_t = \left(\sum_{i=0}^{\infty} \theta_i B^i\right)(1 - \phi B)x_t = a_t,$$

or

$$x_t = (\phi - \theta)\sum_{i=0}^{\infty} \theta^{i-1} x_{t-i} + a_t .$$

The ARMA(1,1) model thus leads to both moving average and autoregressive representations having an infinite number of weights. The ψ-weights converge for $|\phi| < 1$ (the stationarity condition) and the π-weights converge for $|\theta| < 1$ (the invertibility condition). The stationarity condition for the ARMA(1,1) model is thus the same as that of an AR(1) model: the invertibility condition is the same as that of an MA(1) model.

From equation (2.9) it is clear that any product $x_{t-k}a_{t-j}$ has zero expectation if $k > j$. Thus multiplying both sides of (2.8) by x_{t-k} and taking expectations yields

$$\gamma_k = \phi\gamma_{k-1}, \text{ for } k > 1,$$

whilst for $k = 0$ and $k = 1$ we obtain, respectively,

$$\gamma_0 - \phi\gamma_1 = \sigma^2 - \theta(\phi - \theta)\sigma^2 ,$$

and

$$\gamma_1 - \phi\gamma_0 = -\theta\sigma^2 .$$

Eliminating σ^2 from these two equations allows the ACF of the ARMA(1,1) process to be given by

$$\rho_1 = \frac{(1 - \phi\theta)(\phi - \theta)}{1 + \theta^2 - 2\phi\theta} ,$$

and

$$\rho_k = \phi\rho_{k-1}, \text{ for } k > 1 .$$

The ACF of an ARMA(1,1) process is therefore similar to that of an AR(1) process, in that the autocorrelations exponentially decay at a rate ϕ. Unlike the AR(1), however, this decay starts from ρ_1 rather than from $\rho_0 = 1$. Moreover, $\rho_1 \neq \phi$ and, since for typical financial series both ϕ and θ will be positive with $\phi > \theta$, ρ_1 can be much less than ϕ if $\phi - \theta$ is small.

More general ARMA processes are obtained by combining AR(p) and MA(q) processes:

$$x_t - \phi_1 x_{t-1} - \ldots - \phi_p x_{t-p} = a_t - \theta_1 a_{t-1} - \ldots - \theta_q a_{t-q} ,$$

or

$$(1 - \phi_1 B - \ldots - \phi_p B^p) x_t = (1 - \theta_1 B - \ldots - \theta_q B^q) a_t ,$$

i.e.

$$\phi(B) x_t = \theta(B) a_t .$$

The resultant ARMA(p,q) process has the stationarity and invertibility conditions associated with the constituent AR(p) and MA(q) processes respectively. Its ACF will eventually follow the same pattern as that of an AR(p) process after $q-p$ initial values $\rho_1, \ldots, \rho_{q-p}$, while its PACF eventually (for $k>p-q$) behaves like that of an MA(q) process.

Throughout this development, we have assumed that the mean of the process, μ, is zero. Non-zero means are easily accommodated by replacing x_t with $x_t-\mu$, i.e., in the general case of an ARMA(p,q) process, we have

$$\phi(B)(x_t-\mu)=\theta(B)a_t.$$

Noting that $\phi(B)\mu=(1-\phi_1-\ldots-\phi_p)\mu=\phi(1)\mu$, the model can equivalently be written as

$$\phi(B)x_t=\theta_0+\theta(B)a_t,$$

where $\theta_0=\phi(1)\mu$ is a constant or intercept.

2.4 Linear stochastic processes

In the development of ARMA models in the previous section, we assumed that the innovations $\{a_t\}$ were uncorrelated and drawn from a fixed distribution with finite variance, and hence the sequence was termed white noise, i.e., $a_t \sim WN(0,\sigma^2)$. If these innovations are also *independent*, then the sequence $\{a_t\}$ is termed *strict white noise*, denoted $a_t \sim SWN(0,\sigma^2)$, and a stationary process $\{x_t\}$ generated as a linear filter of strict white noise is said to be a *linear* process. It is possible, however, for a linear filter of a white noise process to result in a non-linear stationary process. The distinctions between white and strict white-noise and between linear and non-linear stationary processes are extremely important when modelling financial time series, and will be discussed in much greater detail in chapter 4.

2.5 ARMA model building

2.5.1 Sample autocorrelation and partial autocorrelation functions

An essential first step in fitting ARMA models to observed time series is to obtain estimates of the generally unknown parameters μ, σ_x^2 and the ρ_k. With our stationarity and (implicit) ergodicity assumptions, μ and σ_x^2 can be estimated by the sample mean and sample variance, respectively, of the realisation $\{x_t\}_1^T$

$$\bar{x}=T^{-1}\sum_{t=1}^{T}x_t,$$

$$s^2=T^{-1}\sum_{t=1}^{T}(x_t-\bar{x})^2.$$

An estimate of ρ_k is then given by the lag k *sample autocorrelation*

$$r_k = \frac{\sum_{t=k+1}^{T} (x_t - \bar{x})(x_{t-k} - \bar{x})}{Ts^2}, \quad k = 1,2,\ldots$$

the set of r_ks defining the *sample autocorrelation function* (SACF).

For independent observations drawn from a fixed distribution with finite variance ($\rho_k = 0$, for all $k \neq 0$), the variance of r_k is approximately given by T^{-1} (see, for example, Box and Jenkins, 1976, chapter 2). If, as well, T is large, $\sqrt{T}r_k$ will be approximately standard normal, i.e., $\sqrt{T}r_k \overset{a}{\sim} N(0,1)$, so that an absolute value of r_k in excess of $2T^{-\frac{1}{2}}$ may be regarded as 'significantly' different from zero. More generally, if $\rho_k = 0$ for $k > q$, the variance of r_k, for $k > q$, is

$$V(r_k) = T^{-1}(1 + 2\rho_1^2 + \ldots + 2\rho_q^2). \qquad (2.10)$$

Thus, by successively increasing the value of q and replacing the ρ_js by their sample estimates, the variances of $r_1, r_2, \ldots, r_k$ can be estimated as T^{-1}, $T^{-1}(1 + 2r_1^2), \ldots, T^{-1}(1 + 2r_1^2 + \ldots + 2r_{k-1}^2)$ and, of course, these will be larger, for $k > 1$, than those calculated using the simple formula T^{-1}.

The *sample partial autocorrelation function* (SPACF) is usually calculated by fitting autoregressive models of increasing order: the estimate of the last coefficient in each model is the sample partial autocorrelation, $\hat{\phi}_{kk}$. If the data follow an AR(p) process, then for lags greater than p the variance of $\hat{\phi}_{kk}$ is approximately T^{-1}, so that $\sqrt{T}\hat{\phi}_{kk} \overset{a}{\sim} N(0,1)$.

2.5.2 Model-building procedures

Given the r_k and $\hat{\phi}_{kk}$, with their respective standard errors, the approach to ARMA model building proposed by Box and Jenkins (1976) is essentially to match the behaviour of the SACF and SPACF of a particular time series with that of various theoretical ACFs and PACFs, picking the best match (or set of matches), estimating the unknown model parameters (the ϕ_is, θ_is and σ^2), and checking the residuals from the fitted models for any possible misspecifications.

Another popular method is to select a set of models based on prior considerations of maximum possible settings of p and q, estimate each possible model and select that model which minimises a chosen selection criterion based on goodness of fit considerations. Details of these model-building procedures, and their various modifications, may be found in many texts, e.g., Mills (1990, chapter 8), and hence will not be discussed in detail: rather, they will be illustrated by way of a sequence of examples.

Table 2.1 *SACF of real S&P 500 returns and accompanying statistics*

k	r_k	$s.e.(r_k)$	Q(k)	$Q^*(k)$
1	0.048	0.093	0.27[0.61]	0.27[0.61]
2	−0.160	0.093	3.21[0.20]	3.32[0.19]
3	0.096	0.096	4.28[0.23]	4.44[0.22]
4	−0.040	0.097	4.47[0.35]	4.64[0.33]
5	−0.053	0.097	4.79[0.44]	4.98[0.42]
6	0.014	0.097	4.81[0.57]	5.00[0.54]
7	0.139	0.097	7.04[0.42]	7.42[0.39]
8	−0.109	0.099	8.41[0.39]	8.91[0.35]
9	−0.024	0.100	8.47[0.49]	8.99[0.44]
10	0.051	0.100	8.77[0.55]	9.31[0.50]
11	−0.144	0.102	11.14[0.43]	11.98[0.36]
12	−0.097	0.102	12.22[0.43]	13.21[0.35]

Note: Figures in [..] give $P(\chi^2_k > Q(k), Q^*(k))$

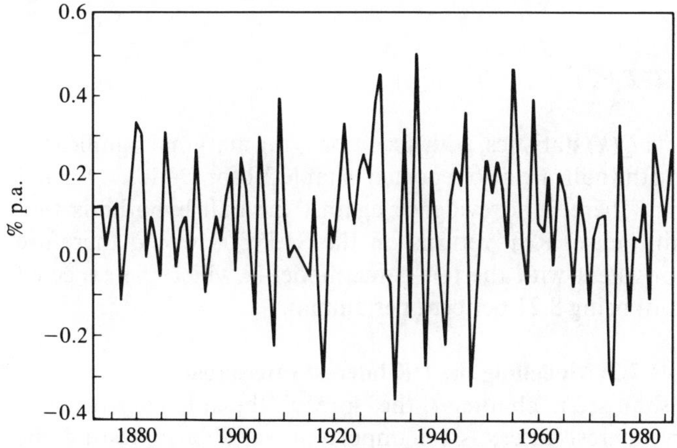

Figure 2.6 S&P 500 index (real returns 1872–1986)

Example 2.1: Are the returns on the S&P 500 a fair game?

An important and often analysed financial series is the real return on the annual Standard & Poor (S&P) 500 stock index for the US. Annual observations from 1872 to 1986 are plotted in figure 2.6 and its SACF up to $k = 12$ is given in table 2.1. The original reference for this data is Shiller (1981a): our series is obtained from the extended data set provided by

Pesaran and Pesaran (1991) as tutorial file SP500.FIT in *MICROFIT 3.0*. It is seen from the plot that the series appears to be stationary about a constant mean, estimated to be 8.21 per cent. This is confirmed by the SACF and a comparison of each of the r_k with their corresponding standard errors, computed using equation (2.10), shows that none are individually significantly different from zero, thus suggesting that the series is, in fact, white noise.

We can also construct a 'portmanteau' statistic based on the complete set of r_ks. On the hypothesis that $x_t \sim WN(\mu,\sigma^2)$, Box and Pierce (1970) show that the statistic

$$Q^*(k) = T\sum_{i=1}^{k} r_i^2$$

is asymptotically distributed as χ^2 with k degrees of freedom, i.e., $Q^*(k) \overset{a}{\sim} \chi_k^2$. Unfortunately, simulations have shown that, even for quite large samples, the true significance levels of $Q^*(k)$ could be much smaller than those given by this asymptotic theory, so that the probability of incorrectly rejecting the null hypothesis is smaller than any chosen significance level. Ljung and Box (1978) argue that a better approximation is obtained when the modified statistic

$$Q(k) = T(T+2)\sum_{i=1}^{k} (T-i)^{-1} r_i^2 \overset{a}{\sim} \chi_k^2$$

is used. $Q^*(k)$ and $Q(k)$ statistics, with accompanying marginal significance levels of rejecting the null, are also reported in table 2.1 for $k = 1, \ldots, 12$, and they confirm that there is no evidence against the null hypothesis that returns are white noise. Real returns on the S&P 500 would therefore appear to be consistent with the fair game model in which the expected return is constant, being 8.21 per cent per annum.

Example 2.2: Modelling the UK interest rate spread

As we shall see in chapter 7, the 'spread', the difference between long and short interest rates, is an important variable in testing the expectations hypothesis of the term structure of interest rates. Figure 2.7 shows the spread between 20 year UK gilts and 91 day Treasury Bills using quarterly observations for the period from 1952 to 1988, while table 2.2 reports the SACF and SPACF up to $k = 12$, with accompanying standard errors. (The spread may be derived from the two interest rate series listed in the data appendix.)

The spread is seen to be considerably smoother than what one would expect if it was a realisation from a white-noise process, and this is confirmed by the SACF, all of whose values are positive, the first five being significantly so (the accompanying portmanteau statistics are computed as

Table 2.2 *SACF and SPACF of the UK spread*

k	r_k	$s.e.(r_k)$	$\hat{\phi}_{kk}$	$s.e.(\hat{\phi}_{kk})$
1	0.829	0.082	0.829	0.082
2	0.672	0.126	−0.048	0.082
3	0.547	0.148	0.009	0.082
4	0.435	0.161	−0.034	0.082
5	0.346	0.169	0.005	0.082
6	0.279	0.174	0.012	0.082
7	0.189	0.176	−0.116	0.082
8	0.154	0.178	0.114	0.082
9	0.145	0.179	0.047	0.082
10	0.164	0.180	0.100	0.082
11	0.185	0.181	0.028	0.082
12	0.207	0.182	0.038	0.082

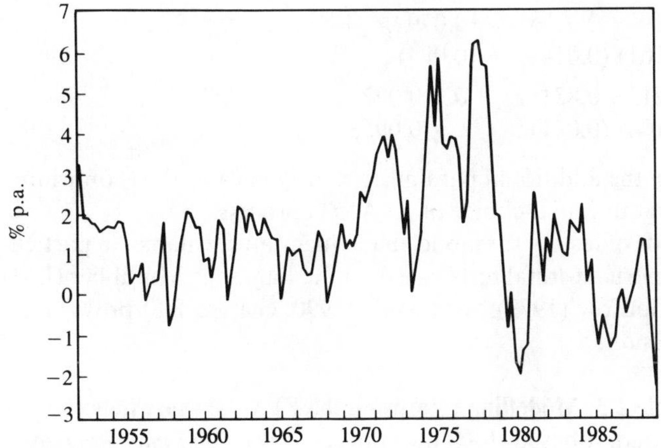

Figure 2.7 UK interest rate spread (quarterly 1952Q1–1988Q4)

$Q^*(12) = 295.2$ and $Q(12) = 305.0$). The SACF in fact displays a roughly exponential decline up till $k = 9$, consistent with there being at least one real root, but it then begins to increase slightly, suggesting the possibility of complex roots. The message from the SPACF is unambiguous, however: with only $\hat{\phi}_{11}$ significant, it identifies an AR(1) process.

Fitting such a process to the series by ordinary least squares (LS) regression yields

$$x_t = \underset{(0.098)}{0.176} + \underset{(0.045)}{0.856} x_{t-1} + \hat{a}_t, \quad \hat{\sigma} = 0.870.$$

Figures in parentheses are standard errors and the intercept implies an estimate of the mean of $\hat{\mu}=\hat{\theta}_0/(1-\hat{\phi}_1)=1.227$, with standard error 0.506.

Having fitted an AR(1) process, it is necessary to check whether such a model is adequate. As a 'diagnostic check', we may examine the properties of the residuals $\hat{a}_t$. Since these are estimates of a_t, they should mimic its behaviour, i.e., they should behave as white noise. The portmanteau statistics Q^* and Q can be used for this purpose, although the degrees of freedom attached to them must be amended: if an ARMA(p,q) process is fitted, then they are reduced to $k-p-q$. With $k=12$, our residuals yield the values $Q^*(12)=12.26$ and $Q(12)=13.21$, now asymptotically distributed as χ^2_{11}, and hence give no evidence of model inadequacy.

An alternative approach to assessing model adequacy is to overfit. For example, we might consider fitting an AR(2) process or, perhaps, an ARMA(1,1), to the series. These yielded the following pair of models (methods of estimating MA processes are discussed, for example, in Mills, 1990, chapter 8.2. We use here conditional LS)

$$\underset{(0.101)}{x_t=0.197}+\underset{(0.054)}{0.927}x_{t-1}-\underset{(0.084)}{0.079}x_{t-2}+\hat{a}_t, \quad \hat{\sigma}=0.869,$$

$$\underset{(0.104)}{x_t=0.213}+\underset{(0.051)}{0.831}x_{t-1}+\hat{a}_t-\underset{(0.095)}{0.092}\hat{a}_{t-1}, \quad \hat{\sigma}=0.870.$$

In both models, the additional parameter is insignificant, thus confirming the adequacy of our initial choice of an AR(1) process.

Other methods of testing for model inadequacy are available. In particular, we may construct formal tests based on the Lagrange Multiplier (LM) principle: see Godfrey (1979), with Mills (1990, chapter 8.8) providing a textbook discussion.

Example 2.3: Modelling returns on the FTA-All share index

The broadest based stock index in the UK is the *Financial Times-Actuaries* (FTA) *All Share*. Table 2.3 reports the SACF and SPACF (up to $k=12$) of its nominal return calculated using monthly observations from 1965 to 1990. Although the portmanteau statistic $Q(12)$ is only 19.3, with a marginal significance level of 0.08, we see that both r_k and $\hat{\phi}_{kk}$ at lags $k=1$ and 3 are greater than two standard errors, thus suggesting that the series is best modelled by some ARMA process of reasonably low order, although a number of processes could be consistent with the behaviour shown by the SACF and SPACF.

In such circumstances, there are a variety of selection criteria that may be used to choose an appropriate model, of which perhaps the most popular is Akaike's (1974) Information Criteria, defined as

$$AIC(p,q)=\ln\hat{\sigma}^2+2(p+q)T^{-1},$$

Table 2.3 *SACF and SPACF of* FTA All Share *Nominal Returns*

k	r_k	$s.e.(r_k)$	$\hat{\phi}_{kk}$	$s.e.(\hat{\phi}_{kk})$
1	0.148	0.057	0.148	0.057
2	−0.061	0.059	0.085	0.057
3	0.117	0.060	0.143	0.057
4	0.067	0.061	0.020	0.057
5	−0.082	0.062	−0.079	0.057
6	0.013	0.062	0.034	0.057
7	0.041	0.063	0.008	0.057
8	−0.011	0.063	0.002	0.057
9	0.087	0.064	0.102	0.057
10	0.021	0.064	−0.030	0.057
11	−0.008	0.064	0.012	0.057
12	0.026	0.064	0.010	0.057

although a criteria that has better theoretical properties is Schwarz's (1978)

$$BIC(p,q)=\ln\hat{\sigma}^2+(p+q)T^{-1}\ln T.$$

A number of other criteria have been proposed, but all are structured in terms of the estimated error variance $\hat{\sigma}^2$ plus a penalty adjustment involving the number of estimated parameters, and it is in the extent of this penalty that the criteria differ. For more discussion about these, and other, selection criteria, see Judge *et al.* (1985, chapter 7.5).

The criteria are used in the following way. Upper bounds, say P and Q, are set for the orders of $\phi(B)$ and $\theta(B)$, and with $\bar{P}=\{0,1,\ldots,P\}$ and $\bar{Q}=\{0,1,\ldots,Q\}$, orders p_1 and q_1 are selected such that, for example

$$AIC(p_1,q_1)=\min AIC(p,q), \quad p\in\bar{P},\, q\in\bar{Q},$$

with parallel strategies obviously being employed in conjunction with *BIC* or any other criterion. One possible difficulty with the application of this strategy is that no specific guidelines on how to determine P and Q seem to be available, although they are tacitly assumed to be sufficiently large for the range of models to contain the true model, which we may denote as having orders (p_0,q_0) and which, of course, will not necessarily be the same as (p_1,q_1), the orders chosen by the criterion under consideration.

Given these alternative criteria, are there reasons for preferring one to another? If the true orders (p_0,q_0) are contained in the set (p,q), $p\in\bar{P}$, $q\in\bar{Q}$, then for all criteria, $p_1\geq p_0$ and $q_1\geq q_0$, almost surely, as $T\rightarrow\infty$. However, *BIC* is *strongly consistent* in that it determines the true model asymptotically, whereas for *AIC* an overparameterised model will emerge no matter how long the available realisation.

Table 2.4 *Model selection criteria for nominal returns*

p \ q	0	1	2	3
AIC				
0	3.701	3.684	3.685	3.688
1	3.689	3.685	3.694	3.696
2	3.691	3.695	3.704	3.699
3	3.683	3.693	3.698	3.707
BIC				
0	3.701	3.696	3.709	3.724
1	3.701	3.709	3.730	3.744
2	3.715	3.731	3.752	3.759
3	3.719	3.741	3.758	3.779

Given the behaviour of the SACF and SPACF of our returns series, we set $\bar{P}=\bar{Q}=3$ and table 2.4 shows the resulting *AIC* and *BIC* values. *AIC* selects the orders (3,0), i.e., an AR(3) process, while *BIC* selects the orders (0,1), so that an MA(1) process is chosen. The two estimated models are

$$\underset{(0.62)}{x_t=4.41}+\underset{(0.057)}{0.173x_{t-1}}-\underset{(0.057)}{0.108x_{t-2}}+\underset{(0.057)}{0.144x_{t-3}}+\hat{a}_t,\quad \hat{\sigma}=6.25,$$

$$x_t=\underset{(0.35)}{5.58}+\hat{a}_t+\underset{(0.056)}{0.186\hat{a}_{t-1}},\quad \hat{\sigma}=6.29.$$

The intercept in the AR(3) model implies a mean estimate of $4.41/(1-0.173+0.108-0.144)=5.58$, exactly the value obtained directly from the MA(1) model, while the first-order MA polynomial $(1+0.186B)$ can be inverted to yield the infinite AR polynomial

$$\begin{aligned}(1+0.186B)^{-1}&=(1-0.186B+0.186^2B^2-0.186^3B^3+\ldots)\\&=(1-0.186B+0.035B^2-0.006B^3),\end{aligned}$$

thus showing that the MA(1) model imposes an exponentially declining autoregressive structure on to the series, rather than a three-period structure. Nevertheless, the sum of these autoregressive coefficients is $(1+0.186)^{-1}=0.843$, which is reasonably close to the sum of the AR(3) coefficients, 0.791. Both models pass diagnostic checks on their residuals, the MA(1) model having a $Q(12)$ value of 15.30, the AR(3) a value of 6.07, with marginal significance levels of 0.23 and 0.91 respectively.

Although, theoretically, the *BIC* has advantages over the *AIC*, it would seem that the latter selects the model that is preferable on more general grounds. However, we should observe that, for both criteria, there are other models that yield criterion values very close to that of the model selected.

Using this idea of being 'close to', Poskitt and Tremayne (1987) introduce the concept of a *model portfolio*. Models are compared to the selected (p_1,q_1) process by way of the statistic, using *AIC* for illustration

$$\mathscr{R}=\exp[-\tfrac{1}{2}T\{AIC(p_1,q_1)-AIC(p,q)\}].$$

Although $\mathscr{R}$ has no physical meaning, its value may be used to 'grade the decisiveness of the evidence' against a particular model. Poskitt and Tremayne (1987) suggest that a value of $\mathscr{R}$ less than $\sqrt{10}$ may be thought of as being a close competitor to (p_1,q_1), with the set of closely competing models being taken as the model portfolio.

Using this concept, with $\sqrt{10}$ taken as an approximate upper bound, we obtain an *AIC* model portfolio containing the six specifications (3,0), (0,1), (0,2), (1,1), (0,3) and (1,0), while the *BIC* portfolio contains the models (0,1), (1,0) and (0,0). We thus see that a wide variety of models appear to offer an adequate description of the returns series.

The 'fair game' model, indexed as (0,0), is contained in the *BIC* model portfolio, although not in that for *AIC*, where higher order models are prevalent, as might be expected from the theoretical properties of this criterion. Nevertheless, the finding of significant third-order processes for returns is plausible, because they almost certainly reflect the seasonality of dividends caused by the tendency of many companies to pay dividends at certain times of the year.

2.6 Non-stationary processes and ARIMA models

The class of ARMA models developed in the previous sections of this chapter relies on the assumption that the underlying process is weakly stationary, thus implying that the mean, variance and autocovariances of the process are invariant under time translations. As we have seen, this restricts the mean and variance to be constant and requires the autocovariances to depend only on the time lag. Many financial time series, however, are certainly not stationary and, in particular, have a tendency to exhibit time-changing means and/or variances.

2.6.1 Non-stationarity in variance

We begin by assuming that a time series can be decomposed into a *non-stochastic* mean level and a random error component

$$x_t=\mu_t+\epsilon_t, \tag{2.11}$$

and we suppose that the variance of the errors, ϵ_t, is functionally related to the mean level μ_t by

$$V(x_t)=V(\epsilon_t)=h^2(\mu_t)\sigma^2,$$

where h is some known function. Our objective is to find a transformation of the data, $g(x_t)$, that will stabilise the variance, i.e., the variance of the transformed variable $g(x_t)$ should be constant. Expanding $g(x_t)$ in a first-order Taylor series around μ_t yields

$$g(x_t) \simeq g(\mu_t) + (x_t - \mu_t)g'(\mu_t),$$

where $g'(\mu_t)$ is the first derivative of $g(x_t)$ evaluated at μ_t. The variance of $g(x_t)$ can then be approximated as

$$\begin{aligned} V[g(x_t)] &\simeq V[g(\mu_t) + (x_t - \mu_t)g'(\mu_t)] \\ &= [g'(\mu_t)]^2 V(x_t) \\ &= [g'(\mu_t)]^2 h^2(\mu_t)\sigma^2. \end{aligned}$$

Thus, in order to stabilise the variance, we have to choose the transformation $g(.)$ such that

$$g'(\mu_t) = \frac{1}{h(\mu_t)}.$$

For example, if the standard deviation of x_t is proportional to its level, $h(\mu_t) = \mu_t$ and the variance-stabilising transformation $g(\mu_t)$ has then to satisfy $g'(\mu_t) = \mu_t^{-1}$. This implies that $g(\mu_t) = \ln\mu_t$, and thus (natural) logarithms of x_t should be used to stabilise the variance. If the variance of x_t is proportional to its level, $h(\mu_t) = \mu_t^{1/2}$, so that $g'(\mu_t) = \mu_t^{-1/2}$. Thus, since $g(\mu_t) = 2\mu_t^{1/2}$, the square root transformation $x_t^{1/2}$ will stabilise the variance. These two examples are special cases of the Box and Cox (1964) class of power transformations

$$g(x_t) = \frac{x_t^\lambda - 1}{\lambda},$$

where we note that $\lim_{\lambda \to 0}[(x_t^\lambda - 1)/\lambda] = \ln x_t$. While the use of logarithms is a popular transformation for financial time series, a constant variance is rarely completely induced by this transformation alone. Chapters 4 and 7 consider various models in which time varying variances are explicitly modelled.

2.6.2 Non-stationarity in mean

A non-constant mean level in equation (2.11) can be modelled in a variety of ways. One possibility is that the mean evolves as a polynomial of order d in time. This will arise if x_t can be decomposed into a trend component, given by the polynomial, and a stochastic, stationary, but possibly autocorrelated, zero mean error component. This is always possible given Cramer's

(1961) extension of Wold's decomposition theorem to non-stationary processes. Thus we may have

$$x_t = \mu_t + \epsilon_t = \sum_{j=0}^{d} \beta_j t^j + \psi(B) a_t. \qquad (2.12)$$

Since

$$E(\epsilon_t) = \psi(B) E(a_t) = 0,$$

we have

$$E(x_t) = E(\mu_t) = \sum_{j=0}^{d} \beta_j t^j,$$

and, as the β_j coefficients remain constant through time, such a trend in the mean is said to be *deterministic*. Trends of this type can be removed by a simple transformation. Consider the linear trend obtained by setting $d = 1$, where, for simplicity, the error component is assumed to be a white-noise sequence

$$x_t = \beta_0 + \beta_1 t + a_t. \qquad (2.13)$$

Lagging (2.13) one period and subtracting this from (2.13) yields

$$x_t - x_{t-1} = \beta_1 + a_t - a_{t-1}. \qquad (2.14)$$

The result is a difference equation following an ARMA(1,1) process in which, since $\phi = \theta = 1$, both autoregressive and moving average roots are unity and the model is neither stationary nor invertible. If we consider the *first differences* of x_t, w_t say, then

$$w_t = x_t - x_{t-1} = (1 - B) x_t = \nabla x_t,$$

where $\nabla = 1 - B$ is known as the *first difference operator*. Equation (2.14) can then be written as

$$w_t = \nabla x_t = \beta_1 + \nabla a_t,$$

and w_t is thus generated by a stationary (since $E(w_t) = \beta_1$ is constant), but not invertible, MA(1) process.

In general, if the trend polynomial is of order d, and ϵ_t is characterised by the ARMA process $\phi(B)\epsilon_t = \theta(B) a_t$, then

$$\nabla^d x_t = (1 - B)^d x_t,$$

obtained by first differencing x_t d times, will follow the process

$$\nabla^d x_t = \theta_0 + \frac{\nabla^d \theta(B)}{\phi(B)} a_t,$$

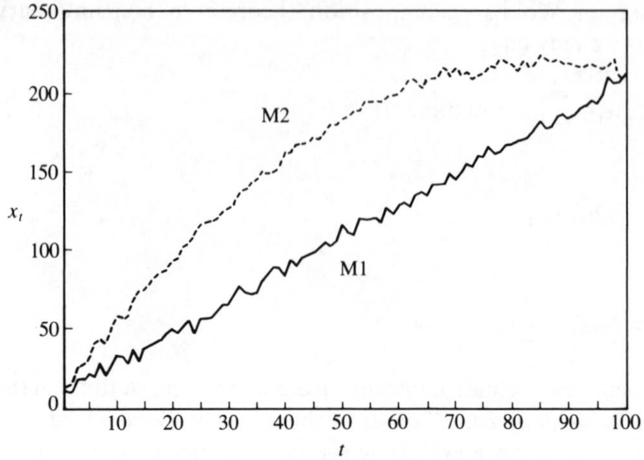

M1: $x_t = 10 + 2t + a_t$; M2: $x_t = 10 + 5t - 0.03t^2 + a_t$
$a_t \sim NID(0, 9)$

Figure 2.8 Linear and quadratic trends

where $\theta_0 = d!\beta_d$. Thus the MA part of the process generating $\nabla^d x_t$ will have d unit roots. Note also that the variance of x_t will be the same as the variance of ϵ_t, which will be constant for all t. Figure 2.8 shows plots of generated data for both linear and quadratic trend models. Because the variance of the error component, here assumed to be white noise and distributed as $NID(0,9)$, is constant and independent of the level, the variability of the two series are bounded about their expected values, and the trend components are clearly observed in the plots.

An alternative way of generating a non-stationary mean level is to consider ARMA models whose autoregressive parameters do not satisfy stationarity conditions. For example, consider the AR(1) process

$$x_t = \phi x_{t-1} + a_t, \tag{2.15}$$

where $\phi > 1$. If the process is assumed to have started at time $t = -N$, the difference equation (2.15) has the solution

$$x_t = x_{-N}\phi^{t+N} + \sum_{i=0}^{t+N} \phi^i a_{t-i}. \tag{2.16}$$

The 'complementary function' $x_{-N}\phi^{t+N}$ can be regarded as the *conditional expectation* of x_t at time $-N$ (Box and Jenkins, 1976, chapter 4), and is an increasing function of t. The conditional expectation of x_t at times $-N+1, -N+2, \ldots, t-2, t-1$ depend on the random shocks a_{-N}, a_{-N+1},

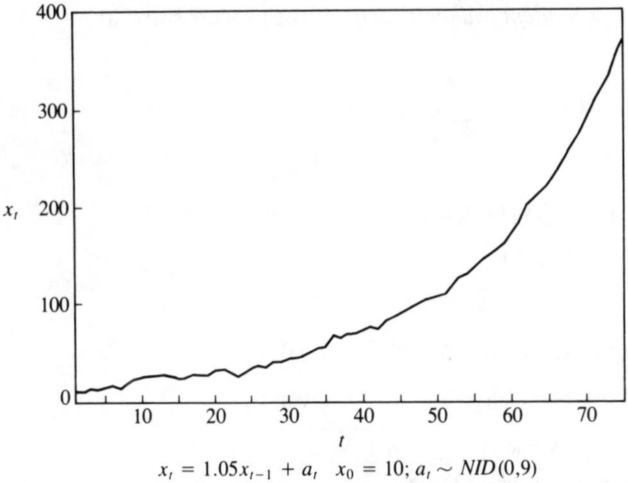

Figure 2.9 Explosive AR(1) model

$\ldots, a_{t-3}, a_{t-2}$, and hence, since this conditional expectation may be regarded as the trend of x_t, the trend changes *stochastically*.

The variance of x_t is given by

$$V(x_t) = \sigma^2 \frac{\phi^{2(t+N+1)} - 1}{\phi^2 - 1},$$

which is an increasing function of time and becomes infinite as $N \to \infty$. In general, x_t will have a trend in both mean and variance, and such processes are said to be *explosive*. A plot of generated data from the process (2.15) with $\phi = 1.05$ and $a_t \sim NID(0,9)$, and having starting value $x_0 = 10$, is shown in figure 2.9. We see that, after a short 'induction period', the series follows essentially an exponential curve with the generating a_ts playing almost no further part. The same behaviour would be observed if further autoregressive and moving average terms were added to the model, as long as the stationarity condition is violated.

As we can see from (2.16), the solution of (2.15) is explosive if $\phi > 1$ but stationary if $\phi < 1$. The case $\phi = 1$ provides a process that is neatly balanced between the two. If x_t is generated by the model

$$x_t = x_{t-1} + a_t, \tag{2.17}$$

then x_t is said to follow a *random walk*. If we allow a constant, θ_0, to be included, so that

$$x_t = x_{t-1} + \theta_0 + a_t, \tag{2.18}$$

then x_t will follow a *random walk with drift*. If the process starts at $t=-N$, then

$$x_t = x_{-N} + (t+N)\theta_0 + \sum_{i=0}^{t+N} a_{t-i},$$

so that

$$\mu_t = E(x_t) = x_{-N} + (t+N)\theta_0,$$
$$\gamma_{0,t} = V(x_t) = (t+N)\sigma^2,$$

and

$$\gamma_{k,t} = Cov(x_t, x_{t-k}) = (t+N-k)\sigma^2, \quad k \geq 0.$$

Thus the correlation between x_t and x_{t-k} is given by

$$\rho_{k,t} = \frac{t+N-k}{\sqrt{(t+N)(t+N-k)}} = \sqrt{\frac{t+N-k}{t+N}}.$$

If $t+N$ is large compared to k, all $\rho_{k,t}$ will be approximately unity. The sequence of x_t values will therefore be very smooth, but will also be non-stationary since both its mean and variance increase with t. Figure 2.10 shows generated plots of the random walks (2.17) and (2.18) with $x_0 = 10$ and $a_t \sim NID(0,9)$. In part (a) of the figure the drift parameter, θ_0, is set to zero while in part (b) we have set $\theta_0 = 2$. The two plots differ considerably, with the drift model showing no affinity whatsoever with the initial value x_0, while the model with $\theta_0 = 0$ only shows some slight affinity. In fact, the expected length of time for a random walk to pass again through an arbitrary value is infinite.

The random walk is an example of a class of non-stationary processes known as *integrated processes*. Equation (2.18) can be written as

$$\nabla x_t = \theta_0 + a_t,$$

and so first differencing x_t leads to a stationary model, in this case the white-noise process a_t. Generally, a series may need first differencing d times to attain stationarity, and the series so obtained may itself be autocorrelated. If this autocorrelation is modelled by an ARMA(p,q) process, then the model for the original series is of the form

$$\phi(B)\nabla^d x_t = \theta_0 + \theta(B)a_t, \tag{2.19}$$

which is said to be an *autoregressive-integrated-moving average* process of orders p, d and q, or ARIMA(p,d,q), and x_t is said to be integrated of order d, denoted $I(d)$. It will usually be the case that the degree of differencing or, equivalently, the order of integration, d, will be 0, 1 or, very occasionally, 2. Again it will be the case that the autocorrelations of an ARIMA process will

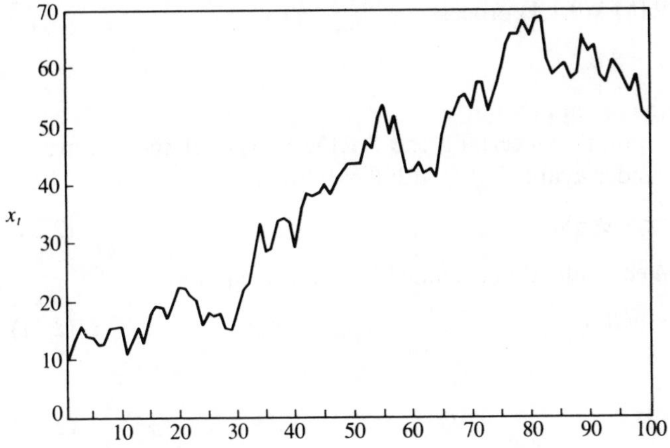

(a) $x_t = x_{t-1} + a_t$, $a_t \sim NID(0, 9)$, $x_0 = 10$

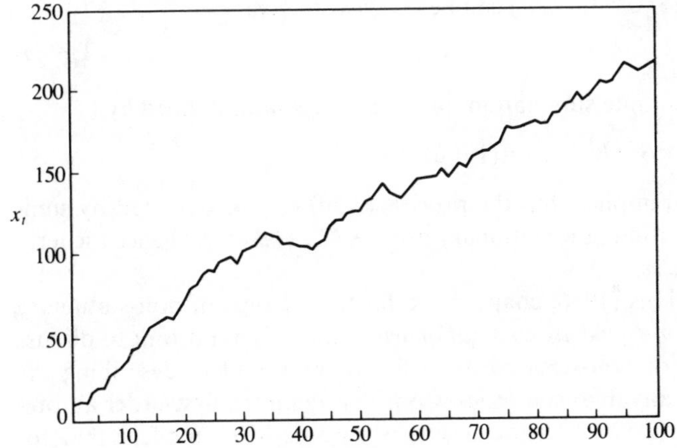

(b) $x_t = 2 + x_{t-1} + a_t$, $a_t \sim NID(0, 9)$, $x_0 = 10$

Figure 2.10 Random walks

be near one for all non-large k. For example, consider the (stationary) ARMA(1,1) process

$$x_t - \phi x_{t-1} = a_t - \theta a_{t-1},$$

whose ACF has been shown to be

$$\rho_1 = \frac{(1-\phi\theta)(\phi-\theta)}{1+\theta^2-2\phi\theta}, \quad \rho_k = \phi\rho_{k-1}, \text{ for } k > 1.$$

As $\phi \to 1$, the ARIMA(0,1,1) process

$$\nabla x_t = a_t - \theta a_{t-1}$$

results, and all the ρ_k tend to unity.

A number of points concerning the ARIMA class of models are of importance. Consider again (2.19), with $\theta_0 = 0$ for simplicity

$$\phi(B)\nabla^d x_t = \theta(B)a_t. \tag{2.20}$$

This process can equivalently be defined by the two equations

$$\phi(B)w_t = \theta(B)a_t, \tag{2.21}$$

and

$$w_t = \nabla^d x_t, \tag{2.22}$$

so that, as we have noted above, the model corresponds to assuming that $\nabla^d x_t$ can be represented by a stationary and invertible ARMA process. Alternatively, for $d \geq 1$, (2.22) can be inverted to give

$$x_t = S^d w_t, \tag{2.23}$$

where S is the infinite summation, or *integral* operator defined by

$$S = (1 + B + B^2 + \ldots) = (1 - B)^{-1} = \nabla^{-1}.$$

Equation (2.23) implies that the process (2.20) can be obtained by summing, or 'integrating', the stationary process (2.21) d times: hence the term integrated process.

Box and Jenkins (1976, chapter 4) refer to this type of non-stationary behaviour as *homogeneous non-stationarity*, and it is important to discuss why this form of non-stationarity is felt to be useful in describing the behaviour of many financial series. Consider again the first-order autoregressive process (2.12). A basic characteristic of the AR(1) model is that, for both $|\phi| < 1$ and $|\phi| > 1$, the local behaviour of a series generated from the model is heavily dependent upon the level of x_t. For many financial series, local behaviour appears to be roughly independent of level, and this is what we mean by homogeneous non-stationarity.

If we want to use ARMA models for which the behaviour of the process is indeed independent of its level, then the autoregressive operator $\phi(B)$ must be chosen so that

$$\phi(B)(x_t + c) = \phi(B)x_t,$$

where c is any constant. Thus

$$\phi(B)c = 0,$$

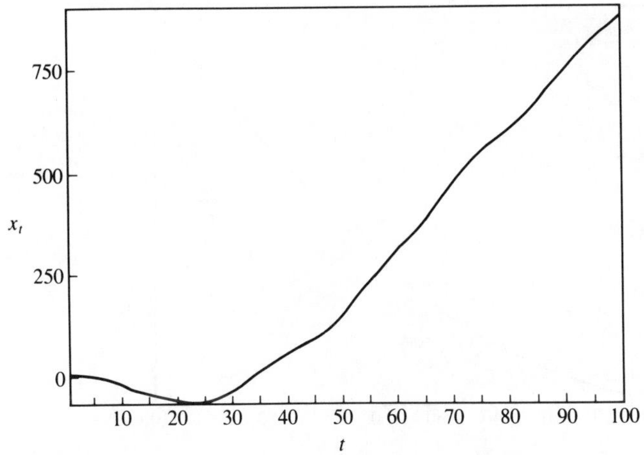

$(1 - B)^2 x_t = a_t,\ x_0 = x_1 = 10,\ a_t \sim NID(0, 2)$

Figure 2.11 'Second difference' model

implying that $\phi(1)=0$, so that $\phi(B)$ must be able to be factorised as

$$\phi(B)=\phi_1(B)(1-B)=\phi_1(B)\nabla ,$$

in which case the class of processes that need to be considered will be of the form

$$\phi_1(B)w_t=\theta(B)a_t ,$$

where $w_t=\nabla x_t$. Since the requirement of homogeneous non-stationarity precludes w_t increasing explosively, either $\phi_1(B)$ is a stationary operator, or $\phi_1(B)=\phi_2(B)(1-B)$, so that $\phi_2(B)w_t^*=\theta(B)a_t$, where $w_t^*=\nabla^2 x_t$. Since this argument can be used recursively, it follows that for time series that are homogeneously non-stationary, the autoregressive operator must be of the form $\phi(B)\nabla^d$, where $\phi(B)$ is a stationary autoregressive operator. Figure 2.11 plots generated data from the model $\nabla^2 x_t=a_t$, where $a_t \sim NID(0,2)$ and $x_0=x_1=10$, and such a series is seen to display random movements in both level and slope.

We see from figures 2.10(a) and 2.11 that ARIMA models without the constant θ_0 in (2.19) are capable of representing series that have *stochastic* trends, which typically will consist of random changes in both the level and slope of the series. As seen from figure 2.10(b) and equation (2.18), however, the inclusion of a non-zero drift parameter introduces a deterministic linear trend into the generated series, since $\mu_t=E(x_t)=\beta_0+\theta_0 t$ if we set $\beta_0=x_{-N}+N\theta_0$. In general, if a constant is included in the model for dth

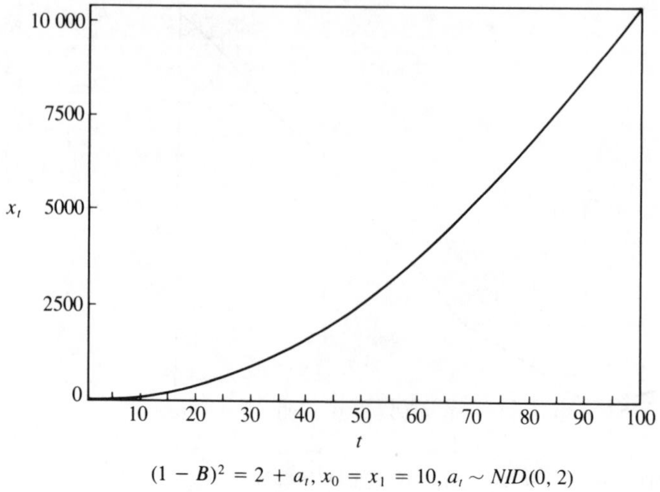

Figure 2.12 'Second difference with drift' model

differences, then a deterministic polynomial trend of degree d is automatically allowed for. Equivalently, if θ_0 is allowed to be non-zero, then

$$E(w_t) = E(\nabla^d x_t) = \mu_w = \theta_0/(1 - \phi_1 - \phi_2 - \ldots - \phi_p)$$

is non-zero, so that an alternative way of expressing (2.19) is as

$$\phi(B)\tilde{w}_t = \theta(B)a_t,$$

where $\tilde{w}_t = w_t - \mu_w$. Figure 2.12 plots generated data for $\nabla^2 x_t = 2 + a_t$, where again $a_t \sim NID(0,2)$ and $x_0 = x_1 = 10$. The inclusion of the deterministic quadratic trend has a dramatic effect on the evolution of the series, with the non-stationary 'noise' being completely swamped after a few periods.

Model (2.19) therefore allows both stochastic and deterministic trends to be modelled. When $\theta_0 = 0$, a stochastic trend is incorporated, while if $\theta_0 \neq 0$, the model may be interpreted as representing a deterministic trend (a polynomial in time of order d) buried in non-stationary noise, which will typically be autocorrelated. The models presented earlier in this section could be described as deterministic trends buried in *stationary* noise, since they can be written as

$$\phi(B)\nabla^d x_t = \phi(1)\beta_d d! + \nabla^d \theta(B)a_t,$$

the stationary nature of the noise in the level of x_t being manifested in d roots of the moving average operator being unity. Further discussion of the

Table 2.5 *SACF and SPACF of the first difference of the UK interest rate spread*

k	r_k	$s.e.(r_k)$	$\hat{\phi}_{kk}$	$s.e.(\hat{\phi}_{kk})$
1	0.002	0.082	0.002	0.082
2	−0.036	0.082	−0.036	0.082
3	−0.040	0.082	−0.040	0.082
4	−0.090	0.082	−0.092	0.082
5	−0.090	0.084	−0.094	0.082
6	0.063	0.085	0.054	0.082
7	−0.192	0.086	−0.211	0.082
8	−0.068	0.092	−0.085	0.082
9	−0.101	0.093	−0.143	0.082
10	−0.038	0.094	−0.076	0.082
11	−0.002	0.094	−0.065	0.082
12	0.173	0.094	0.097	0.082

relationships between stochastic and deterministic trends is contained in chapter 3.

2.7 ARIMA modelling

Once the order of differencing d has been established then, since $w_t = \nabla^d x_t$ is by definition stationary, the ARMA techniques discussed in section 2.5.2 may be applied to the suitably differenced series. Establishing the correct order of differencing is by no means straightforward, however, and is discussed in detail in chapter 3. We content ourselves here with a sequence of examples illustrating the modelling of ARIMA processes when d has already been chosen: the suitability of these selections are examined through examples in the subsequent chapter.

Example 2.4: Modelling the UK spread as an integrated process

In example 2.2 we modelled the spread of UK interest rates as a stationary, indeed AR(1), process. Here we consider modelling the spread assuming that it is an $I(1)$ process, i.e., we examine the behaviour of the SACF and SPACF of $w_t = \nabla x_t$. Table 2.5 provides these estimates up to $k = 12$ and shows that there is no evidence against the hypothesis that w_t is white noise (the accompanying portmanteau statistic is $Q(12) = 16.74$). Indeed, since the mean of w_t is not significantly different from zero, having a point estimate of -0.04 with an accompanying standard error of 0.07, this implies that the spread can be modelled as a driftless random walk.

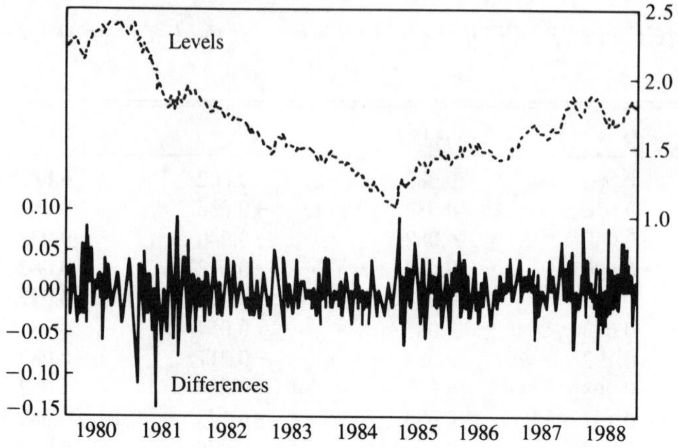

Figure 2.13 Dollar/sterling exchange rate (weekly 1980–8)

Example 2.5: Modelling the dollar/sterling exchange rate

Figure 2.13 plots weekly observations of both the level and first differences of the dollar/sterling exchange rate from January 1980 to December 1988, a total of 470 observations. The levels exhibit the wandering movement of a driftless random walk and, consistent with this, the differences are stationary about zero and show no discernable pattern, appearing to be a realisation from a white-noise process. This is confirmed by examination of the SACF and SPACF of the differences: none of the first twenty-four sample autocorrelations or partial autocorrelations are significantly different from zero, having an accompanying portmanteau statistic of $Q(24)=20.3$. Moreover, the sample mean is estimated as -0.0009, with standard error 0.0013, thus confirming the absence of drift in the series.

Example 2.6: Modelling the *FTA All Share* index

Figure 2.14 plots monthly observations from January 1965 to December 1990 of the *FTA All Share* index and, as expected, shows the series to exhibit a prominent upward, but not linear, trend, with pronounced and persistent fluctuations about it, which increase in variability as the level of the series increases. This behaviour thus suggests a logarithmic transformation to be appropriate. The so transformed observations are also shown in figure 2.14: taking logarithms does indeed both linearise the trend and stabilise the variance.

Eliminating the trend by taking first differences yields the SACF and

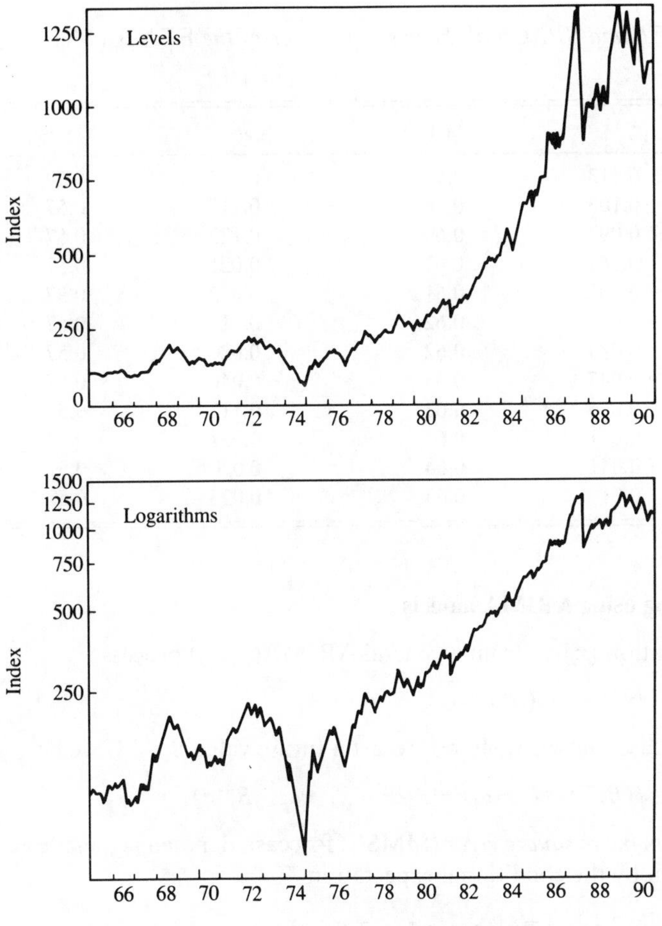

Figure 2.14 *FTA All Share* index (monthly 1965–90)

SPACF shown in table 2.6. Although the r_ks show no discernible pattern, the $\hat{\phi}_{kk}$s suggest an AR(3) process, whose adequacy is confirmed through overfitting and residual diagnostic tests. The fitted model is

$$\underset{}{\nabla x_t} = \underset{(0.0038)}{0.0067} + \underset{(0.057)}{0.141}\nabla x_{t-1} - \underset{(0.057)}{0.133}\nabla x_{t-2} + \underset{(0.057)}{0.123}\nabla x_{t-3} + \hat{a}_t, \quad \hat{\sigma} = 0.0646, \quad Q(12) = 7.62$$

The implied estimate of μ is 0.0077 which, since ∇x_t can be interpreted as the monthly growth of the index, implies an annual mean growth rate of approximately 9.25 per cent. Additional interpretation of this model, particularly in the forecasting context, is provided in example 2.7.

Table 2.6 *SACF and SPACF of the first difference of the* FTA All Share Index

k	r_k	$s.e.(r_k)$	$\hat{\phi}_{kk}$	$s.e.(\hat{\phi}_{kk})$
1	0.113	0.57	0.113	0.57
2	−0.103	0.58	−0.117	0.57
3	0.093	0.60	0.122	0.57
4	0.061	0.61	0.021	0.57
5	−0.102	0.61	−0.092	0.57
6	−0.036	0.62	−0.011	0.57
7	0.044	0.62	0.020	0.57
8	−0.047	0.63	−0.046	0.57
9	0.075	0.63	0.115	0.57
10	0.021	0.64	−0.031	0.57
11	−0.041	0.64	−0.020	0.57
12	0.019	0.64	0.023	0.57

2.8 Forecasting using ARIMA models

Given a realisation $\{x_t\}_{1-d}^{T}$ from a general ARIMA(p,d,q) process

$$\phi(B)\nabla^d x_t = \theta_0 + \theta(B)a_t,$$

it is often the case that we wish to forecast a future value x_{T+h}. If we let

$$\alpha(B) = \phi(B)\nabla^d = (1 - \alpha_1 B - \alpha_2 B^2 - \ldots - \alpha_{p+d}B^{p+d}),$$

then a *minimum mean square error* (MMSE) forecast, denoted $f_{T,h}$, made at time T, is given by the conditional expectation

$$f_{T,h} = E(\alpha_1 x_{T+h-1} + \alpha_2 x_{T+h-2} + \ldots + \alpha_{p+d}x_{T+h-p-d} + \theta_0 + a_{T+h} - \theta_1 a_{T+h-1} - \ldots - \theta_q a_{T+h-q} \mid x_T, x_{T-1}, \ldots).$$

Now

$$E(x_{T+j} \mid x_T, x_{T-1}, \ldots) = \begin{cases} x_{T+j}, & j \leq 0 \\ f_{T,j}, & j > 0 \end{cases}$$

and

$$E(a_{T+j} \mid x_T, x_{T-1}, \ldots) = \begin{cases} a_{T+j}, & j \leq 0 \\ 0 & j > 0 \end{cases}$$

so that, to evaluate $f_{T,h}$, all we need to do is: (i) replace past expectations ($j \leq 0$) by known values, x_{T+j} and a_{T+j}, and (ii) replace future expectations ($j > 0$) by forecast values, $f_{T,j}$ and 0.

Three examples will illustrate the procedure. Consider first the AR(1) model $(1-\phi B)x_t=\theta_0+a_t$, so that $\alpha(B)=(1-\phi B)$. Here

$$x_{T+h}=\phi x_{T+h-1}+\theta_0+a_{T+h},$$

and hence

$$\begin{aligned} f_{T,h}&=\phi f_{T,h-1}+\theta_0 \\ &=\phi^h x_T+\theta_0(1+\phi+\phi^2+\ldots+\phi^{h-1}) \end{aligned}$$

by repeated substitution. Thus, for stationary processes ($|\phi|<1$), as $h\rightarrow\infty$,

$$f_{T,h}=\frac{\theta_0}{1-\phi}=E(x_t)=\mu,$$

so that for large lead times the best forecast of a future observation is eventually the mean of the process.

Next consider the ARIMA(0,1,1) model $\nabla x_t=(1-\theta B)a_t$. Here $\alpha(B)=(1-B)$ and so

$$x_{T+h}=x_{T+h-1}+a_{T+h}-\theta a_{T+h-1}.$$

For $h=1$, we have

$$f_{T,1}=x_T-\theta a_T,$$

for $h=2$,

$$f_{T,2}=f_{T,1}=x_T-\theta a_T.$$

and, in general,

$$f_{T,h}=f_{T,h-1},\quad h>1.$$

Thus, for all lead times, the forecasts from origin T will follow a straight line parallel to the time axis passing through $f_{T,1}$. Note that, since

$$f_{T,h}=x_T-\theta a_T$$

and

$$a_T=(1-B)(1-\theta B)^{-1}x_T,$$

the h-step ahead forecast can be written as

$$\begin{aligned} f_{T,h}&=(1-\theta)(1-\theta B)^{-1}x_T \\ &=(1-\theta)(x_T+\theta x_{T-1}+\theta^2 x_{T-2}+\ldots), \end{aligned}$$

i.e., the forecast for all future values of x is an exponentially weighted moving average of current and past values.

Finally, consider the ARIMA(0,2,2) model $\nabla^2 x_t=(1-\theta_1 B-\theta_2 B^2)a_t$, with $\alpha(B)=(1-B)^2=(1-2B+B^2)$

$$x_{T+h}=2x_{T+h-1}-x_{T+h-2}+a_{T+h}-\theta_1 a_{T+h-1}-\theta_2 a_{T+h-2}.$$

For $h=1$, we have

$$f_{T,1}=2x_T-x_{T-1}-\theta_1 a_T-\theta_2 a_{T-1},$$

for $h=2$,

$$f_{T,2}=2f_{T,1}-x_T-\theta_2 a_T,$$

for $h=3$,

$$f_{T,3}=2f_{T,2}-f_{T,1},$$

and thus, for $h\geq 3$

$$f_{T,h}=2f_{T,h-1}-f_{T,h-2}.$$

Hence, for all lead times, the forecasts from origin n will follow a straight line passing through the forecasts $f_{T,1}$ and $f_{T,2}$.

The h-step ahead forecast error for origin T is

$$e_{T,h}=x_{T+h}-f_{T,h}=a_{T+h}+\psi_1 a_{T+h-1}+\ldots+\psi_{h-1}a_{T+1},$$

where $\psi_1, \ldots, \psi_{h-1}$ are the first $h-1$ ψ-weights in $\psi(B)=\alpha^{-1}(B)\theta(B)$. The variance of this forecast error is then

$$V(e_{T,h})=\sigma^2(1+\psi_1^2+\psi_2^2+\ldots+\psi_{h-1}^2). \qquad (2.24)$$

The forecast error is therefore a linear combination of the unobservable future shocks entering the system after time T and, in particular, the one-step ahead forecast error ($h=1$) is

$$e_{T,1}=x_{T,1}-f_{T,1}=a_{T+1}.$$

Thus, for a MMSE forecast, the one-step ahead forecast errors must be uncorrelated. However, it is not the case that h-step ahead forecasts made at different origins will be uncorrelated, nor will be forecasts for different lead times made at the same origin (see, for example, Box and Jenkins, 1976, appendix A5.1).

For the AR(1) example given above, since $\psi_j=\phi^j$, the variance of the h-step ahead forecast is given by

$$V(e_{T,h})=\sigma^2(1+\phi^2+\phi^4+\ldots+\phi^{2(h-1)})$$
$$=\sigma^2\frac{(1-\phi^{2h})}{(1-\phi^2)}.$$

Thus, as h tends to infinity, the variance increases to a constant value $\sigma^2/(1-\phi^2)$, this being the variation of the process about the ultimate forecast, μ.

For the ARIMA(0,1,1) model, $\psi_j=1-\theta$, $(j=1,2,\ldots)$. Thus we have

$$V(e_{T,h}) = \sigma^2(1 + (h-1)(1-\theta)^2),$$

which increases with h. Similarly, the ARIMA(0,2,2) model has ψ-weights given by $\psi_j = 1 + \theta_2 + j(1 - \theta_1 - \theta_2)$, $(j = 1,2,\ldots)$, and an h-step ahead forecast error variance of

$$V(e_{T,h}) = \sigma^2(1 + (h-1)(1+\theta_2)^2 + \tfrac{1}{6}h(h-1)(2h-1)(1-\theta_1-\theta_2)^2 + h(h-1)(1+\theta_2)(1-\theta_1-\theta_2)),$$

which again increases with h.

The examples in this section thus show how the degree of differencing, or order of integration, determines not only how successive forecasts are related to each other, but also the behaviour of the associated forecast error variances.

Example 2.7: ARIMA forecasting of financial time series

Here we examine the properties of ARIMA forecasts for some of the series analysed in the examples of this chapter.

Example 2.1 fitted an AR(1) model to the UK interest rate spread, yielding parameter estimates $\hat{\phi} = 0.856$, $\hat{\theta}_0 = 0.176$ and $\hat{\sigma} = 0.870$. With the end-of-sample spread being $x_T = -2.27$, h-step ahead forecasts are provided by the formula

$$f_{T+h} = 0.856^h(-2.27) + 0.176(1 + 0.856 + 0.856^2 + \ldots + 0.856^{h-1}),$$

which tend, eventually, to 1.23, the sample mean of the spread. These point forecasts are accompanied by forecast error variances given by

$$V(e_{T,h}) = 0.870^2(1 + 0.856^2 + 0.856^4 + \ldots + 0.856^{2(h-1)}),$$

which tend to the constant $0.870^2/(1-0.856^2) = 2.832$.

If, however, we use the random walk process of example 2.4 to model the spread then, using the results for the ARIMA(0,1,1) process with $\theta = 0$ and $\hat{\sigma} = 0.894$, the estimate of σ obtained from this model, our forecasts are simply $f_{T+h} = -2.27$ for *all h*, and there is thus *no* tendency for the forecasts to converge to the sample mean or, indeed, to any other value. Furthermore, the forecast error variances are given by $V(e_{T,h}) = \sigma^2 h = 0.80h$ and hence *increase* with h, rather than tending to a constant. This example thus illustrates within the forecasting context the radically different properties of autoregressive models which have, on the one hand, a unit root and, on the other, a root that is large but less than unity.

The dollar/sterling exchange rate was also found, in example 2.5, to follow a random walk, which thus implies that, given the end-of-sample exchange rate of 1.81, all future forecasts of the rate are that particular value, although the precision of the forecasts, given by the accompanying forecast error variance, diminishes as the forecast horizon increases: with σ estimated to be 0.027, we have $V(e_{T,h})=0.0073h$.

In example 2.6 we modelled the logarithms of the *FTA All Share* index as an ARIMA(3,1,0) process. Since

$$\phi(B)=(1-0.141B+0.133B^2-0.123B^3),$$

we have

$$\alpha(B)=(1-1.141B+0.274B^2-0.256B^3+0.123B^4),$$

so that forecasts can be computed recursively as

$$\begin{aligned}
f_{T,1} &= 1.141x_T - 0.274x_{T-1} + 0.256x_{T-2} - 0.123x_{T-3} + 0.0067,\\
f_{T,2} &= 1.141f_{T,1} - 0.274x_T + 0.256x_{T-1} - 0.123x_{T-2} + 0.0067,\\
f_{T,3} &= 1.141f_{T,2} - 0.274f_{T,1} + 0.256x_T - 0.123x_{T-1} + 0.0067,\\
f_{T,4} &= 1.141f_{T,3} - 0.274f_{T,2} + 0.256f_{T,1} - 0.123x_T + 0.0067,
\end{aligned}$$

and, for $h\geq 5$

$$f_{T,h}=1.141f_{T,h-1}-0.274f_{T,h-2}+0.256f_{T,h-3}-0.123f_{T,h-4}+0.0067.$$

By computing the coefficients in the polynomial $\psi(B)=\alpha^{-1}(B)$ as

$$\psi(B)=1+1.141B+1.028B^2+1.229B^3+1.392B^4+\ldots \qquad (2.25)$$

and using the estimate $\hat{\sigma}=0.646$, forecast error variances can then be computed using the formula (2.24): since the series is $I(1)$, these variances increase with h.

Additional interpretation of the nature of these forecasts is provided by the *eventual forecast function*, which is obtained by solving the difference equation implicit in the ARIMA(3,1,0) representation of x_t at time $T+h$ (see, for example, Mills, 1990, chapter 7.3, for a general development and McKenzie, 1988, for further discussion)

$$\begin{aligned}
&x_{T+h}-1.141x_{T+h-1}+0.274x_{T+h-2}-0.256x_{T+h-3}+0.123x_{T+h-4}\\
&=0.0067+a_{T+h}.
\end{aligned}$$

At origin T, this difference equation has the solution

$$x_{T+h}=\sum_{i=1}^{4}b_i^{(T)}f_i(h)+0.0067\sum_{j=T+1}^{T+h}\psi_{T+h-j},$$

where the ψs are as in (2.25) and the functions $f_1(h),\ldots,f_4(h)$ depend upon the roots of the polynomial $\alpha(B)$, which are unity, a real root of 0.45, and a pair of complex roots $(0.15\pm0.50i)$. Hence, the solution can be written as

$$\begin{aligned}
x_{T+h}&=b_0+b_1^{(T)}+b_2^{(T)}(0.45)^h\\
&\quad+b_3^{(T)}(0.15+0.50i)^h+b_4^{(T)}(0.15-0.50i)^h,
\end{aligned}$$

where

$$b_0 = 0.0067 \sum_{j=T+1}^{T+h} \psi_{T+h-j}.$$

For a given origin T, the coefficients $b_j^{(T)}, j = 1, \ldots, 4$, are constants applying to all lead times h, but they change from one origin to the next, adapting themselves to the observed values of x_t. They can be obtained by solving a set of recursive equations containing the $f_i(h)$s, ψ_h and a_T.

Since the ψs increase with h, b_0 imparts a deterministic drift into x_{T+h}, so that $b_0 + b_1^{(T)}$ gives the forecasted 'trend' of the series. Around this trend are a geometrically declining component, provided by the real, but stationary, root, and a damped sine wave provided by the pair of complex roots, its damping factor, frequency and phase being functions of the process parameters (Box and Jenkins, 1976, pages 58–63).

3 Univariate linear stochastic models: further topics

The previous chapter has demonstrated that the order of integration, *d*, is a crucial determinant of the properties that a time series exhibits. This chapter begins with an exposition of the techniques available for determining the order of integration of a time series, emphasising the importance of the chosen alternative hypothesis to the null of a unit root: in particular, whether the alternative is that of a constant mean, a linear trend, or a segmented trend. The importance of these models to finance is demonstrated through a sequence of examples.

We then move, in section 2, to examining methods of decomposing an observed time series into two or more unobserved components, emphasising the signal extraction approach to estimating these components. This approach is particularly suited to estimating, under assumptions of market efficiency, expected, *ex ante*, values using only observed, *ex post*, observations and is illustrated by showing how expected real interest rates can be extracted from observed rates.

The final sections of the chapter focus attention on long-term properties of financial time series. A number of recent models of stock market behaviour yield the prediction that stock returns, far from being unpredictable, should exhibit negative autocorrelation over long time horizons: that they should be *mean reverting*. Section 3 thus develops techniques for measuring and testing for such mean reversion, or *persistence* as it is often referred to. Section 4 introduces an alternative method of modelling long-term memory in a time series, through the use of a *fractional* value of *d*. Fractionally integrated extensions of ARIMA models are developed and methods of testing for fractional integration and estimating such models are introduced. Both methods of modelling long-run behaviour are illustrated through examples using stock returns.

3.1 Determining the order of integration of a time series

3.1.1 Unit root tests

As stated above, the order of integration, d, is a crucial determinant of the properties that a time series exhibits. If we restrict ourselves to the most common values of zero or one for d, so that x_t is either $I(0)$ or $I(1)$, then it is useful to bring together the properties of such processes.

If x_t is $I(0)$, which we will denote $x_t \sim I(0)$ even though such a notation has been used previously to denote the distributional characteristics of a series, then, assuming for convenience that it has zero mean,

(i) the variance of x_t is finite and does not depend on t,
(ii) an innovation a_t has only a temporary effect on the value of x_t,
(iii) the expected length of times between crossings of $x=0$ is finite, i.e., x_t fluctuates around its mean of zero,
(iv) the autocorrelations, ρ_k, decrease steadily in magnitude for large enough k, so that their sum is finite.

If $x_t \sim I(1)$ with $x_0=0$, then

(i) the variance of x_t goes to infinity as t goes to infinity,
(ii) an innovation a_t has a permanent effect on the value of x_t because x_t is the sum of all previous innovations: see, e.g., equation (2.16),
(iii) the expected time between crossings of $x=0$ is infinite,
(iv) the autocorrelations $\rho_k \rightarrow 1$ for all k as t goes to infinity.

The fact that a time series is non-stationary is often self evident from a plot of the series. Determining the actual form of non-stationarity, however, is not so easy from just a visual inspection, and an examination of the SACFs for various differences may be required.

To see why this may be so, recall that a stationary AR(p) process requires that all the roots g_i in

$$\phi(B)=(1-g_1B)(1-g_2B)\ldots(1-g_pB)$$

are such that $|g_i|<1$. Now suppose that one of them, say g_1, approaches 1, i.e., $g_1=1-\delta$, where δ is a small positive number. The autocorrelations

$$\rho_k = A_1g_1^k + A_2g_2^k + \ldots + A_pg_p^k \simeq A_1g_1^k$$

will then be dominated by $A_1g_1^k$, since all other terms will go to zero more rapidly. Furthermore, as g_1 is close to 1, the exponential decay $A_1g_1^k$ will be slow and almost linear, since

$$A_1g_1^k = A_1(1-\delta)^k = A_1(1-\delta k+\delta^2k^2-\ldots) \simeq A_1(1-\delta k).$$

Hence, failure of the SACF to die down quickly is therefore an indication of non-stationarity, its behaviour tending to be that of a slow, linear decline. It should also be pointed out that this slow decay can start at values of r_1 considerably smaller than 1, sometimes 0.5 or even less. If the original series x_t is found to be non-stationary, the first difference ∇x_t is then analysed. If ∇x_t is still non-stationary, the second difference $\nabla^2 x_t$ is analysed: the procedure being repeated until a stationary difference is found, although it is seldom the case in practice that d exceeds 2.

Sole reliance on the SACF can sometimes lead to problems of *overdifferencing*. Although further differences of a stationary series will themselves be stationary, overdifferencing can lead to serious difficulties. Consider the stationary MA(1) process $x_t = (1-\theta B)a_t$. The first difference of this is

$$\begin{aligned}\nabla x_t &= (1-B)(1-\theta B)a_t \\ &= (1-(1+\theta)B+\theta B^2)a_t \\ &= (1-\theta_1 B-\theta_2 B^2)a_t.\end{aligned}$$

We now have a more complicated model containing two parameters rather than one and, moreover, one of the roots of the $\theta(B)$ polynomial is unity since $\theta_1+\theta_2=1$. The model is therefore not invertible, so that the AR(∞) representation does not exist and attempts to estimate this model will almost surely run into difficulties.

Note also that the variance of x_t is given by

$$V(x) = \gamma_0(x) = (1+\theta^2)\sigma^2,$$

whereas the variance of $w_t = \nabla x_t$ is given by

$$\begin{aligned}V(w) = \gamma_0(w) &= (1+(1+\theta)^2+\theta^2)\sigma^2 \\ &= 2(1+\theta+\theta^2)\sigma^2.\end{aligned}$$

Hence

$$V(w) - V(x) = (1+\theta)^2\sigma^2 > 0,$$

thus showing that the variance of the overdifferenced process will be larger than that of the original MA(1) process. The behaviour of the sample variances associated with different values of d can provide a useful means of deciding the appropriate level of differencing: the sample variances will decrease until a stationary sequence has been found, but will tend to increase on overdifferencing. However, this is not always the case, and a comparison of sample variances for successive differences of a series is best employed as a useful auxiliary method of determining the appropriate value of d.

Both the above approaches are informal methods based on descriptive, sample statistics and it is legitimate to ask whether there are more formal tests available by which we may determine the value of d. More precisely, we wish to be able to test for the presence of one or more *unit roots* in the pth order autoregressive polynomial $\phi(B)$ in the model

$$\phi(B)x_t = \theta_0 + \theta(B)a_t, \tag{3.1}$$

where x_0 is assumed to be fixed and $\theta_0 = \phi(1)\mu$, μ being the mean of x_t.

The theory and practice of testing for unit roots has produced a voluminous literature in recent years, and this has been formally reviewed in Fuller (1985), Perron (1988) and Diebold and Nerlove (1990), and conveniently surveyed for applied researchers by Dickey, Bell and Miller (1986) and Dolado, Jenkinson and Sosvilla-Rivero (1990). To develop this testing procedure, let us concentrate on autoregressive models, which may not be particularly restrictive since (3.1) will always have an autoregressive representation if, as assumed, the moving average polynomial $\theta(B)$ is invertible. Equation (3.1) can be written as

$$x_t = \theta_0 + \rho_1 x_{t-1} - \sum_{j=1}^{p-1} \rho_{j+1} \nabla x_{t-j} + a_t, \tag{3.2}$$

where

$$\rho_i = \sum_{j=1}^{p} \phi_j, \quad i = 1, \ldots, p.$$

Since $\phi(B)$ will contain a unit root if $\sum_1^p \phi_j = 1$, the presence of such a root is formally equivalent to the null hypothesis $\rho_1 = 1$.

If p is known, the coefficients in (3.2) can be consistently estimated by LS regression and the coefficient on x_{t-1} thus provides a means of testing the null hypothesis $\rho_1 = 1$ against the stationary alternative $\rho_1 < 1$. An obvious test statistic is the usual 't-ratio' of the estimate of $(\rho_1 - 1)$ to its estimated standard error. Dickey and Fuller (1979), however, show that this statistic does not have a Student's t distribution: the distribution, denoted τ_μ to distinguish it from the conventional t, has selected percentiles published in, for example, Fuller (1976, p. 373). To appreciate the differences in the two distributions, these percentiles show that for large sample sizes, using a 0.05 significance level would require a critical τ_μ value of -2.86, rather than -1.645 for the normal approximation to Student's t. It should be pointed out, though, that the distribution tabulated by Dickey and Fuller explicitly assumes that, as well as x_0 being fixed at zero, the intercept θ_0 under the null hypothesis is zero. Although this will be formally the case in (3.2) since $\phi(1) = 0$ under the null, it may not be the case in practice and Schmidt (1990) tabulates amendments to the τ_μ distribution when there is such 'drift' under the null hypothesis.

Note that an equivalent, and somewhat more convenient, form for the test is given by the t-ratio on x_{t-1} in the regression

$$\nabla x_t = \theta_0 + (\rho_1 - 1)x_{t-1} + \sum_{j=1}^{p-1} \rho_{j+1} \nabla x_{t-j} + a_t. \tag{3.3}$$

If x_t is generated by an ARIMA(p,1,q) process, with both p and q assumed known, then Said and Dickey (1985) show that the τ_μ statistic calculated from the regression

$$\nabla x_t = \theta_0 + (\rho_1 - 1)x_{t-1} + \sum_{j=1}^{p-1} \rho_{j+1} \nabla x_{t-j} + a_t - \sum_{j=1}^{q} \theta_j a_{t-j} \tag{3.4}$$

has the same asymptotic distribution as that calculated from (3.3).

The problem here is that p and q are assumed known, and this is unlikely to be the case in practice. When p and q are unknown, Said and Dickey (1984) show that, under the null hypothesis of a unit root, the test statistic obtained from equation (3.3) can still be used if the number of lags of ∇x_t introduced as regressors increases with the sample size T at a controlled rate ($T^{\frac{1}{3}}$). Schwert (1987) suggests setting the number of lags, k, equal to either $[4(T/100)^{0.25}]$ or $[12(T/100)^{0.25}]$, where [.] denotes the operation of taking the integer part of the argument, depending upon the amount and seasonal nature of the data available, while Diebold and Nerlove (1990) find that $[T^{0.25}]$ works well in practice. The adjustment is necessary because as the sample size increases, the effects of the correlation structure of the residuals on the shape of the distribution of τ_μ become more precise.

The assumption made concerning the innovation sequence a_t in the development of this test is the usual one that the innovations are a white-noise sequence having constant variance. With financial time series, it is often the case that an allowance for heterogeneity of variance is desirable. Phillips (1987a, 1987b, see also Phillips and Perron, 1988) relaxes the assumptions on the sequence $\{a_t\}$ to allow a wide range of serial correlation and heterogeneity patterns (these conditions are discussed in greater detail in chapter 6), and considers the test statistic for a unit root when $p=1$ in (3.3), i.e.,

$$\nabla x_t = \theta_0 + (\rho_1 - 1)x_{t-1} + a_t. \tag{3.5}$$

Since no lags of ∇x_t are included as additional regressors, under these more general conditions on $\{a_t\}$ τ_μ requires 'correcting' for any general forms of serial correlation and/or heterogeneity that might be present. Phillips proposes adjusting the statistic to

$$Z(\tau_\mu) = \tau_\mu(\hat{\sigma}/\hat{\sigma}_{\tau l}) - \tfrac{1}{2}(\hat{\sigma}^2_{\tau l} - \hat{\sigma}^2)T\left\{\hat{\sigma}^2_{\tau l} \sum_{t=2}^{T} (x_{t-1} - \bar{x}_{-1})^2\right\}^{-\frac{1}{2}},$$

where $\hat{\sigma}^2$ is the sample variance of a_t,

$$\bar{x}_{-1}=(T-1)^{-1}\sum_{t=1}^{T-1}x_t,$$

and

$$\hat{\sigma}^2_{\tau l}=T^{-1}\sum_{t=1}^{T}\hat{a}_t^2+2T^{-1}\sum_{j=1}^{l}\omega_{jl}\sum_{t=j+1}^{T}\hat{a}_t\hat{a}_{t-j},$$

the $\hat{a}_t$ being the residuals from the regression of (3.5). The triangular set of weights $\omega_{jl}=\{1-j/(l+1)\}$ ensure that the variance estimate $\hat{\sigma}^2_{\tau l}$ is positive (Newey and West, 1987): l can be chosen using one of the rules suggested by Schwert (1987) above. The $Z(\tau_\mu)$ statistic can again be compared to the τ_μ tables.

Phillips (1987a) shows that this 'non-parametric' test has the same *asymptotic* power under general error structures as the 'parametric' τ_μ statistic. However, its finite sample properties appear to be markedly worse than the τ_μ statistic when the moving average structure in (3.4) has a root close to minus unity, although when the root is positive, e.g., if $q=1$ and $\theta_1<0$, then there are potential gains in using $Z(\tau_\mu)$ rather than τ_μ. Further discussion of the power of unit root tests is contained in the following subsection.

As we have seen in chapter 1, the presence of a unit root is often a theoretical implication of models which postulate the rational use of information available to economic agents, and thus unit roots occur in many theoretical financial models: for example, variables such as future contracts and stock prices (Samuelson, 1965, 1973), dividends and earnings (Kleidon, 1986a), spot and forward exchange rates (Meese and Singleton, 1982), and interest rates (Campbell and Shiller, 1987) should all contain unit roots under rational expectations. Unit root tests are thus extremely important in the analysis of financial time series.

Example 3.1: Unit root tests on financial time series

Examples 2.2 and 2.4 examined two models for the UK interest rate spread, an AR(1) process and a driftless random walk; while example 2.7 compared and contrasted the two models. We are now in a position to discriminate between the two through the application of a unit root test. The fitted AR(1) model

$$\underset{(0.098)}{x_t=0.176}+\underset{(0.045)}{0.856}x_{t-1}+a_t$$

can be interpreted as the unit root test equation (3.2) with $p=1$. With $\hat{\rho}_1=0.856$, our test statistic is calculated as $\tau_\mu=(0.856-1)/0.045=-3.20$.

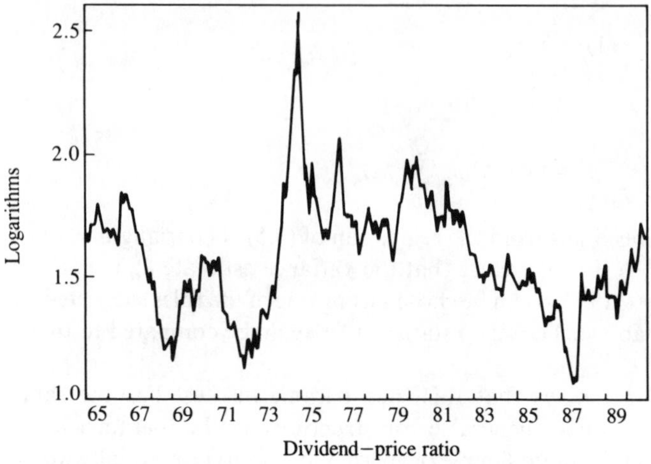

Figure 3.1 *FTA All Share* index (monthly 1965–90)

Using MacKinnon's (1990) extended tabulations of critical values, this value is sufficient to reject the null hypothesis of a unit root in favour of the alternative of a root less than unity at the 0.05 significance level ($\tau_{\mu,0.05} = -2.88$), and we may thus conclude that the appropriate model for the spread is a stationary AR(1) process.

We should also note that the estimate of the innovation standard deviation σ under the null hypothesis is 0.894, rather higher than its estimate from the AR(1) model ($\hat{\sigma} = 0.870$), both of which are reported in example 2.7. This is thus in accordance with the above discussion on overdifferencing.

A similar approach to testing for a unit root in the dollar/sterling exchange rate, the presence of which was assumed in example 2.5, leads to the estimated regression

$$\underset{}{\nabla x_t = \underset{(0.0066)}{0.0103} - \underset{(0.0039)}{0.0067} x_{t-1},}$$

and here the τ_μ statistic, calculated to be -1.72, is clearly insignificant ($\tau_{\mu,0.10} = -2.57$ in this case). The exchange rate is thus confirmed as a random walk.

Figure 3.1 plots the logarithm of the dividend–price ratio of the UK *FTA All Share* index for the period January 1965 to December 1990. Although the series does not contain a trend, its wandering pattern could be a consequence of it being generated by an $I(1)$ process and hence a unit root test may be performed. Examination of the SACFs of both the levels and the differences showed that autocorrelation was present, so that $p > 1$ and lags of ∇x_t need to be included as auxiliary regressors in the unit root regression. Since $T=308$, setting $k=[T^{0.25}]$ necessitates introducing three lags into the regression. Such a regression provides an estimate of ρ_1 of -0.955, having a standard error of 0.016. The resulting τ_μ statistic of -2.89 is thus (just) significant at the 0.05 level ($\tau_{\mu,0.05}$ here being equal to -2.87), so that we reject the null of a unit root in favour of the alternative that the log dividend–price ratio is stationary, a finding that will be shown in chapter 7 to be consistent with the implications of one of the most fundamental models in finance: that of the present value relationship linking real stock prices and dividends.

3.1.2 Trend stationarity versus difference stationarity

In the unit root testing strategy just outlined, the implicit null hypothesis is that the series is generated as a driftless random walk with, possibly, a serially correlated error: in the terminology of Nelson and Plosser (1982), x_t is said to be *difference stationary* (DS)

$$\nabla x_t = \epsilon_t, \tag{3.6}$$

where $\epsilon_t = \theta(B)a_t$, while the alternative is that x_t is *stationary* in levels. While the null of a driftless random walk is appropriate for many financial series such as interest rates and exchange rates, other series often do contain a drift, so that the relevant null becomes

$$\nabla x_t = \theta + \epsilon_t. \tag{3.7}$$

In this case, a plausible alternative is that x_t is generated by a linear trend buried in stationary noise (see chapter 2.6), i.e., it is *trend stationary* (TS)

$$x_t = \beta_0 + \beta_1 t + \epsilon_t. \tag{3.8}$$

Perron (1988) shows that neither the τ_μ statistic obtained from (3.4) nor its non-parametric counterpart $Z(\tau_\mu)$ are capable of distinguishing a stationary process around a linear trend (model (3.8)) from a process with a unit root and drift (model (3.7)). Indeed, rejection of the null hypothesis of a unit root is unlikely if the series is stationary around a linear trend and becomes impossible as the sample size increases.

A test of (3.7) against (3.8) is, however, straightforward to carry out by using an extension of the testing methodology discussed above. If the parametric testing procedure is used, then the 't-ratio' on x_{t-1} from the regression

$$\nabla x_t = \beta_0 + \beta_1 t + (\rho_1 - 1)x_{t-1} + \sum_{i=1}^{k} \delta_i \nabla x_{t-i} + a_t \tag{3.9}$$

is the appropriate test statistic, and can be compared with the tables of the τ_τ statistic given originally in Fuller (1976) and extended to a much wider range of sample sizes and critical values by Guilkey and Schmidt (1989) and MacKinnon (1990). Note that the presence of the time trend t as an additional regressor alters the distribution of the statistic (hence the change in subscript from μ to τ) and thus the critical values: for example, in large samples the 0.05 critical value is $\tau_\tau = -3.41$ as opposed to the τ_μ value of -2.86 given earlier. If the non-parametric approach is employed, then a time trend regressor may be added to (3.5) and the analogous adjusted t-ratio can again be compared to the τ_τ tables. The adjustment in this case is

$$Z(\tau_\tau) = \tau_\tau(\hat{\sigma}/\hat{\sigma}_{\tau l}) - (\hat{\sigma}_{\tau l}^2 - \hat{\sigma}^2)T^3(4\hat{\sigma}_{\tau l}[3D_{xx}]^{\frac{1}{2}})^{-1},$$

where

$$D_{xx} = (T^2(T^2-1)/12)\sum x_{t-1}^2 - T(\sum tx_{t-1})^2 + T(T+1)\sum tx_{t-1}\sum x_{t-1} - (T(T+1)(2T+1)/6)(\sum x_{t-1})^2$$

is the determinant of the regressor moment matrix of the time trend augmented regression (3.5).

Strictly, the unit root null requires not only that $\rho_1 = 1$ in (3.9) but also that $\beta_1 = 0$, because if $\beta_1 \neq 0$, x_t will contain a *quadratic* trend. Nelson and Plosser (1982) argue that this is not usually a problem, since we would typically wish to preclude the possibility that the process has a unit root *and* a trend a priori: if x_t is in logarithms, a non-zero β_1 under the null implies an ever increasing (or decreasing) rate of change ∇x_t.

A parametric joint test of $\rho_1 = 1$ and $\beta_1 = 0$ is given by the usual regression 'F-test' comparing the residual sum of squares from the regression of (3.9) with the residual sum of squares from the 'restricted' regression of ∇x_t on an intercept and k lags of ∇x_t. The resulting statistic, denoted Φ, is then compared with the table of critical values presented in Dickey and Fuller (1981). A non-parametric counterpart, $Z(\Phi)$, is also available: its formula is given by Perron (1988).

Unfortunately, the τ_τ and $Z(\tau_\tau)$ statistics are not invariant with respect to β_1, even asymptotically. If $\beta_1 \neq 0$ under the null, the variance of x_t is dominated by a quadratic trend and τ_τ will be asymptotically normal. As we have remarked earlier, a similar situation arises with τ_μ and $Z(\tau_\mu)$: if $\beta_0 \neq 0$ (with $\beta_1 = 0$) under the null, so that the variance of x_t is dominated by a linear trend, τ_μ will be asymptotically normal.

To circumvent such complications, Dolado, Jenkinson and Sosvilla-Rivero (1990) propose the following strategy for testing for unit roots in the presence of possible trends. Equation (3.9) is first estimated and τ_τ used to test the null hypothesis that $\rho_1 = 1$. If the null is rejected, there is no need to go further and the testing procedure stops. If the null is not rejected, we test for the significance of β_1 under the null, i.e., we estimate

$$\nabla x_t = \beta_0 + \beta_1 t + \sum_{i=1}^{k} \delta_i \nabla x_{t-i} + a_t \tag{3.10}$$

and test whether β_1 is zero or not using conventional testing procedures. If β_1 is significant, we compare τ_τ with the standardised normal and make our inference on the null accordingly. If, on the other hand, β_1 is not significant, we estimate (3.9) without the trend ($\beta_1 = 0$)

$$\nabla x_t = \beta_0 + (\rho_1 - 1)x_{t-1} + \sum_{i=1}^{k} \delta_i \nabla x_{t-i} + a_t \tag{3.11}$$

and test the unit root null of $\rho_1 = 1$ using τ_μ. If the null is rejected, again the testing procedure is terminated. If it is not rejected, we test for the significance of the constant β_0 under the null using the regression

$$\nabla x_t = \beta_0 + \sum_{i=1}^{k} \delta_i \nabla x_{t-i} + a_t . \tag{3.12}$$

If β_0 is insignificant, we conclude that x_t contains a unit root, while if $\beta_0 \neq 0$, we compare τ_μ with the standardised normal, again making our inference accordingly.

This procedure is, of course, based on the *asymptotic* normality of τ_τ and τ_μ in the presence of a trend or drift in the relevant unit root null. An interesting question is what happens in small samples? Both Hylleberg and Mizon (1989a) and Schmidt (1990) present evidence that, when the drift parameter β_0 is small compared to σ^2, the small sample distribution of τ_μ is very much closer to the Dickey–Fuller distribution than to the standard normal. Indeed, Schmidt presents tabulations of the τ_μ distribution for various values of the 'standardised drift' β_0/σ_a, although the accuracy of these tabulations when the drift is estimated has yet to be established. On the basis of this evidence, then, we should be careful when applying the asymptotic normality result in the testing strategy outlined above.

Various other approaches to testing for unit roots have been proposed: Hall's (1989) instrumental variable method for the case when x_t contains a moving average component (see also Pantula and Hall, 1991, for an ARMA extension), Bhargava's (1986) most powerful invariant tests when $k = 0$ in (3.9), and Phillips and Ouliaris' (1988) technique of using the 'long-run' variance of ∇x_t are all summarised in Dolado, Jenkinson and Sosvilla-Rivero (1990), while Park (1990) provides a survey of the 'variable addition' class of unit root tests.

A related development is the set of confidence intervals for the largest autoregressive root provided by Stock (1991). Stock assumes that the true value of ρ_1 can be modelled as $\rho_1 = 1 + c/T$, where c is a fixed constant, and uses 'local-to-unity' asymptotic distribution theory to construct asymptotic confidence intervals for ρ_1 based on computed τ_μ and τ_τ statistics. Since the distributions of these statistics are non-normal and the dependence on c is not a simple location shift, such confidence intervals cannot be constructed using a simple rule such as '± 2 standard errors'. The intervals are highly non-linear, exhibiting a sharp 'bend' for c just above zero (see Stock, 1991, figures 1 and 2): for positive values of the test statistics the intervals are tight, for large negative values they are wide.

Stock provides tables from which confidence intervals for ρ_1 can be calculated given a value of τ_μ or τ_τ and the sample size T. As an illustration of such a calculation, recall that the τ_μ statistic for the UK interest rate spread was reported in example 3.1 to be -3.20. From Part A of table A.1 in Stock (1991), such a value corresponds to a 95 per cent confidence interval for c of $(-30.93, -1.57)$. Since the statistic was computed from a sample of size $T = 148$, this corresponds to an interval for ρ_1 of $(1-30.93/148,\ 1-1.57/148)$, i.e., $0.791 \leq \rho_1 \leq 0.989$. Since $\hat{\rho}_1 = 0.856$, this shows clearly the complicated nature of the relation between $\hat{\rho}_1$ and the confidence interval constructed by 'inverting' the τ statistic: the point estimate is not, and generally will not be, at the centre of the interval. Nevertheless, unity is excluded from the interval, thus confirming our choice of a stationary process for modelling this series.

This discussion of unit root testing would not be complete without some remarks about power. Unlike many hypothesis testing situations, the power of tests of the unit root hypothesis against stationary alternatives depends very little on the number of observations per se and very much on the *span* of the data. For a given number of observations, the power is largest when the span is longest. Conversely, for a given span, additional observations obtained using data sampled more frequently lead only to a marginal increase in power, the increase becoming negligible as the sampling interval is decreased: see Shiller and Perron (1985) and Perron (1991). Hence a data set containing fewer annual observations over a long time period will lead to unit root tests having higher power than those computed from a data set containing more observations over a shorter time period. This is of some importance when analysing financial time series, which very often have a large number of observations obtained by sampling at very fine intervals over a fairly short time span: in such circumstances, the unit root null may be very difficult to reject even if false, particularly when the alternative is a root that is close to, but less than, unity. Monte Carlo evidence supporting this line of argument is provided by DeJong *et al.* (1992a, 1992b).

Related to this point is Cochrane's (1991a) observation that any TS process can be approximated arbitrarily well by a unit root process, and vice versa, in the sense that the ACFs of the two processes will be arbitrarily close. This can be illustrated in a straightforward way by considering the ARMA(1,1) process

$$x_t = \phi x_{t-1} + a_t - \theta a_{t-1}.$$

This process is DS when $\phi = 1$ and $-1 < \theta < 1$, but TS (with a zero trend) when $-1 < \phi < 1$ and $-1 < \theta < 1$. Consider the case where the TS process has $\phi = \theta = 0$ (so the series is white noise), while the DS process has $\phi = 1$ and $0 < \theta < 1$. For any finite sample size, the TS process will be approximated arbitrarily well by the DS process, in the sense defined above, provided θ is close enough to, but not equal to, unity. Campbell and Perron (1991) point out that this implies that, in finite samples, any test of the unit root hypothesis against TS alternatives must have power no greater than its size, and vice versa, so that trying to discriminate between the two might well be regarded as impossible.

They then go on to argue that, nevertheless, distinguishing between the two classes of process can yield important advantages, particularly when forecasting (near-integrated stationary processes seem to be better forecasted using integrated models, while near-stationary unit root processes appear to be better forecasted using stationary models) and when finite sample distributions need to be approximated by either stationary or integrated asymptotic distributions: it may be better to use integrated asymptotic theory for near-integrated stationary models and stationary asymptotic theory for near-stationary integrated models.

We should emphasise, however, that all these testing procedures rely on classical methods of statistical inference. Recently, a Bayesian approach to the testing of unit roots has been developed: see De Jong and Whiteman (1991a, 1991b), Sims (1988), Sims and Uhlig (1991) and Koop (1992). This has provoked a fierce interchange between Phillips (1991a, 1991b) and various Bayesian critics, the whole of the October–December 1991 issue of *The Journal of Applied Econometrics* being devoted to the subject. It is clear that this is a methodological issue that will take some time to resolve (if at all!): for the present it would seem that the classical approach developed above remains the most convenient for the applied practitioner to adopt.

Example 3.2: Are UK equity prices trend or difference stationary?

In example 2.6 we modelled the logarithms of the UK *FTA All Share* index as an ARIMA(3,1,0) process on noting that it had a pronounced tendency to drift upwards, albeit with some major 'wanderings' about trend. We may thus test the null hypothesis that the series contains a unit root against the alternative that it is trend stationary, i.e., generated as stationary deviations about a linear trend.

Following the testing strategy outlined above requires estimating the following regressions (with absolute t-ratios now shown in parentheses)

$$\underset{}{\nabla x_t} = \underset{(2.31)}{0.124} + \underset{(2.29)}{0.00025t} - \underset{(2.27)}{0.0284x_{t-1}} + \sum_{i=1}^{3}\hat{\delta}_i \nabla x_{t-i} + \hat{a}_t, \tag{i}$$

$$\nabla x_t = -\underset{(0.27)}{0.0023} + \underset{(1.18)}{0.00007t} + \sum_{i=1}^{3}\hat{\delta}_i \nabla x_{t-i} + \hat{a}_t, \tag{ii}$$

$$\nabla x_t = \underset{(0.63)}{0.0166} - \underset{(0.38)}{0.00176x_{t-1}} + \sum_{i=1}^{3}\hat{\delta}_i \nabla x_{t-i} + \hat{a}_t, \tag{iii}$$

$$\nabla x_t = \underset{(1.78)}{0.0067} + \sum_{i=1}^{3}\hat{\delta}_i \nabla x_{t-i} + \hat{a}_t. \tag{iv}$$

From regression (i) a τ_τ test cannot reject the DS null, but β_1 is found to be insignificant under this null from regression (ii). This necessitates estimating regression (iii), from which a τ_μ test still cannot reject the null. Estimating equation (iv) shows that β_0 is non-zero under the null (using a 0.05 significance level on a one-sided test), so that τ_μ strictly should be tested against a standard normal. Since $\tau_\mu = -0.38$, however, this does not alter our conclusion that x_t is a DS process: equity prices thus follow an $I(1)$ process with drift, so that the model estimated in example 2.6 is the appropriate one. Note that the implied estimate of ρ_1 from regression (i) is 0.972 and, with $T = 308$, the associated confidence interval, calculated using Part B of table A.1 of Stock (1991), is $0.948 \leq \rho_1 \leq 1.014$.

3.1.3 Segmented trends and structural breaks

The null hypothesis so far considered is that the observed series $\{x_t\}_0^T$ is a realisation of a process characterised by the presence of a unit root and possibly a non-zero drift. Perron (1989a) generalises the approach so as to allow a one-time change in the structure occurring at a time $T_B (1 < T_B < T)$. He considers three different models under the null hypothesis: one that permits an exogenous change in the level of the series (a 'crash'), one that permits an exogenous change in the rate of growth, and one that allows both changes. The hypotheses are parameterised as follows:

Null hypotheses:

Model (A) $x_t = \mu + x_{t-1} + dD(TB)_t + e_t$,
Model (B) $x_t = \mu_1 + x_{t-1} + (\mu_2 - \mu_1)DU_t + e_t$,
Model (C) $x_t = \mu_1 + x_{t-1} + dD(TB)_t + (\mu_2 - \mu_1)DU_t + e_t$,

where

$$D(TB)_t = 1 \quad \text{if } t = T_B + 1, \quad 0 \text{ otherwise;}$$
$$DU_t = 1 \quad \text{if } t > T_B, \quad 0 \text{ otherwise;}$$

and where $\{e_t\}$ is an error process having the properties of the error in equation (3.5). Perron considers the following alternative hypotheses:

Alternative hypotheses:

Model (A) $x_t = \mu_1 + \beta t + (\mu_2 - \mu_1)DU_t + e_t,$
Model (B) $x_t = \mu + \beta_1 t + (\beta_2 - \beta_1)DT_t^* + e_t,$
Model (C) $x_t = \mu_1 + \beta_1 t + (\mu_2 - \mu_1)DU_t + (\beta_2 - \beta_1)DT_t^* + e_t,$

where $DT_t^* = t - T_B$ if $t > T_B$ and 0 otherwise. T_B is the time of the break, i.e., the period at which the change in the parameters of the trend function occurs.

Model (A), the 'crash model', characterises the null hypothesis of a unit root by a dummy variable which takes the value one at the time of the break. Under the alternative hypothesis of a trend stationary system, Model (A) allows for a one-time change in the intercept of the trend function, the magnitude of this change being $\mu_2 - \mu_1$. Model (B), the 'changing growth' model, specifies under the null that the drift parameter changes from μ_1 to μ_2 at time T_B, while under the alternative, a change in the slope of the trend function (of magnitude $\beta_2 - \beta_1$), without any sudden change in the level, is allowed (a segmented trend). Model (C) allows both effects to take place simultaneously, i.e., a sudden change in the level followed by a different growth path.

Perron (1989a) shows that standard tests of the unit root hypothesis are not consistent against 'trend stationary' alternatives when the trend function contains a shift in the slope (see also Rappoport and Reichlin, 1989, and Reichlin, 1989). Although such tests are not inconsistent against a shift in the intercept of the trend function (if the change is fixed as T increases), their power is likely to be substantially reduced due to the fact that the limit of the autoregressive coefficient is inflated above its true value.

Perron (1989a) thus extends the unit root testing strategy to ensure a consistent testing procedure against shifting trend functions. Rather than use the observed series $\{x_t\}$ to construct test statistics, we use the 'detrended' series, $\{\tilde{x}_t^i\}$, whose form depends on which model is being considered. Thus we let $\{\tilde{x}_t^i\}$, $i = A, B, C$, be the residuals from a regression of x_t on:

(1) $i = A$: a constant, a time trend, and DU_t;
(2) $i = B$: a constant, a time trend, and DT_t^*;
(3) $i = C$: a constant, a time trend, DU_t, and DT_t^*.

Test statistics can then be constructed in the usual way, but the standard Fuller (1976) percentiles are now inappropriate. Perron (1989a) provides the appropriate tables, which now depend not on T but on the time of the break relative to the sample size, i.e., on the ratio T_B/T. In general, these percentiles are larger (in absolute value) than their Fuller counterparts, with the largest value occurring when the break is in mid sample ($T_B/T=0.5$).

One drawback of these models is that they imply that the change in the trend function occurs instantaneously. Perron (1989a) generalises the models by assuming that x_t responds to a shock in the trend function in the same way as it reacts to any other shock. Concentrating attention on the most general case, that of a sudden change in level followed by a different growth path (Model C), the procedure is to estimate the regression

$$x_t = \mu + \theta DU_t + \beta t + \gamma DT_t^* + dD(TB)_t + \alpha x_{t-1} + \sum_{i=1}^{k} c_i \nabla x_{t-i} + e_t.$$

Under the null hypothesis of a unit root, $\alpha=1$ and $\theta=\beta=\gamma=0$, whereas under the alternative of a 'trend stationary' process, $\alpha<1$ and θ, β, $\gamma\neq 0$. Moreover, under the null d should be non-zero while under the alternative it should be close to zero.

Rappoport and Reichlin (1989) extend these tests to the situation where there are two posited breaks of trend, although their simulation experiments only provide a limited set of critical values. The case when there is no trend in the model is considered in Perron (1989b).

A major problem with testing for unit roots in the suspected presence of a structural break is that, as Christiano (1988) and Zivot and Andrews (1990) demonstrate, the choice of the timing of the break (or set of breaks) is unlikely to be selected exogenously: in practice, a break point will be decided upon by a combination of visual examination of data plots, consultation with colleagues, and formal testing techniques. This break point is then tested for formal significance. Christiano shows that whether or not the computed significance level takes into account pre-test examination of the data can make a drastic difference: the true significance levels are usually very much higher than the nominal ones that assume that the break point was chosen exogenously.

Zivot and Andrews (1990), on the other hand, consider the null hypothesis to be simply that $\{x_t\}$ is $I(1)$ without an exogenous structural break and view the selection of the break point $\lambda=T_B/T$ as the outcome of an estimation procedure designed to fit $\{x_t\}$ to a certain TS representation, i.e., they assume that the alternative hypothesis stipulates that $\{x_t\}$ can be represented by a TS process with a single break in trend occurring at an unknown point in time. Their approach is to choose the λ which minimises the unit root test statistic computed from the regression

$$x_t=\mu+\theta DU_t(\lambda)+\beta t+\gamma DT_t^*(\lambda)+\alpha x_{t-1}+\sum_{i=1}^{k} c_i\nabla x_{t-i}+e_t,$$

where $DU_t(\lambda)=1$ and $DT_t^*(\lambda)=t-[T\lambda]$ if $t>[T\lambda]$, and 0 otherwise, the regression being estimated with the break fraction λ ranging from $2/T$ to $(T-1)/T$. Note that, under the null that x_t is $I(1)$ with no structural break, the dummy $D(TB_t)$ is no longer needed, and Zivot and Andrews allow the number of lagged ∇x_ts to vary with λ.

When the selection of λ is treated as the outcome of an estimation procedure, Perron's critical values can no longer be used: they are too small in absolute value and hence are biased towards rejecting the unit root null. Zivot and Andrews (1990, tables 2A–4B) provide critical values for the limiting distributions of these 'minimum t-ratios', along with recomputed critical values for the distributions when λ is regarded as fixed. The difference between the critical values is substantial: for λ fixed at 0.5, the 0.05 critical value is -4.24, whereas the corresponding value for λ estimated is -5.08.

A related approach to the problem of distinguishing between integrated and segmented trend models is provided by Balke and Fomby (1991a, 1991b), who use intervention analysis to identify outliers caused by infrequent permanent shocks (see Mills, 1990, chapter 12, for a textbook discussion of such techniques).

Example 3.3: Unit roots and structural breaks in US stock prices

Figure 3.2 plots the logarithms of the nominal annual US S&P stock index for the period 1871 to 1988, the data source being the same as the real return series analysed in example 2.1. A conventional unit root test obtained a value of $\tau_\tau=-1.84$ which, since $\tau_{\tau,0.10}=-3.15$, thus provides no evidence to reject the null hypothesis that stock prices are DS in favour of the alternative that they are TS. Following Perron (1989a), however, we first consider the possibility of both a change in level and, thereafter, an increased trend rate of growth of the series in the wake of the Great Crash of 1929. We thus set the break point T_B at 1929 and estimate the regression

$$\begin{aligned} x_t=&\underset{(0.117)}{0.570}+\underset{(0.0017)}{0.0070t}-\underset{(0.068)}{0.216DU_t}+\underset{(0.003)}{0.011DT_t^*}-\underset{(0.185)}{0.177DTB_t}\\ &+\underset{(0.062)}{0.707x_{t-1}}+\underset{(0.092)}{0.156\nabla x_{t-1}},\end{aligned}$$

where now standard errors are shown in parentheses. The t-ratio for testing $\alpha=1$ yields a value of -4.74 and, since T_B/T is approximately 0.5, this is significant at the 0.05 level. Moreover, since θ, β and γ are all significant and

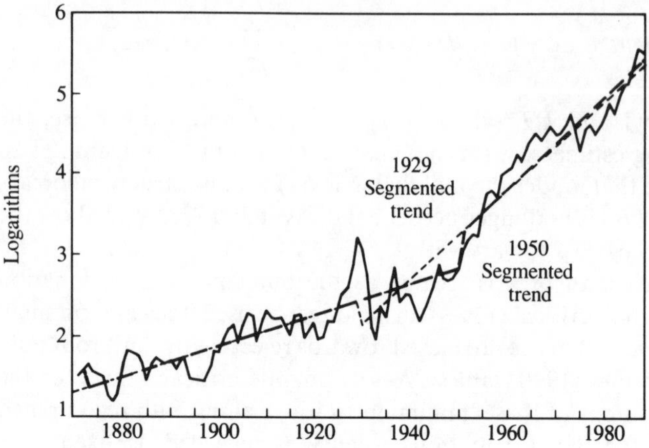

Figure 3.2 Nominal S&P 500 index (1970–88)

d is insignificantly different from zero, all findings are thus consistent with x_t being generated by a segmented trend process. Indeed, the following model can be obtained

$$\underset{}{x_t} = \underset{(0.085)}{1.859} + \underset{(0.0047)}{0.0188t} + \underset{(0.0071)}{0.0366DT_t^*} - \underset{(0.152)}{0.312DU_t} + u_t$$

$$u_t = \underset{(0.093)}{0.932u_{t-1}} - \underset{(0.094)}{0.200u_{t-2}} + e_t, \quad \hat{\sigma}_e = 0.1678.$$

The trend function from this model is superimposed on the stock price series in figure 3.2: prices grew at a trend rate of 1.88 per cent per annum up to 1929 and 5.55 per cent thereafter, the crash provoking a decrease of almost 25 per cent in the trend level of the series.

However, we may not wish to impose the break point onto the series and might prefer to estimate it using the procedure proposed by Zivot and Andrews (1990). This led to the break point being estimated as $T_B = 1950$, with an associated test statistic of -5.09, which is just significant at the 0.05 level. The implied segmented trend model is estimated as

$$x_t = \underset{(0.064)}{1.871} + \underset{(0.0024)}{0.0193t} + \underset{(0.0069)}{0.0391DT_t^*(\hat{\lambda})} + \underset{(0.144)}{0.377DU_t(\hat{\lambda})} + u_t$$

$$u_t = \underset{(0.094)}{0.869u_{t-1}} - \underset{(0.093)}{0.201u_{t-2}} + e_t, \quad \hat{\sigma}_e = 0.1638.$$

This model, which produces a slightly better fit, implies that prices grew at a trend rate of 1.93 per cent per annum until 1950 and 5.84 per cent thereafter, the shock in that year increasing the series by over 50 per cent. This trend is also superimposed on the price series in figure 3.2: the similarity of the two segmented trends is readily apparent, the question is whether there is a crash in 1929 or an upward jump in 1950!

Whichever model is preferred, the finding that stock prices can be modelled as a segmented trend is, as Perron (1989a, p. 1384) remarks, 'particularly striking given the vast amount of theoretical and empirical studies supporting the random walk hypothesis in this situation'.

3.1.4 Testing for more than one unit root

The above development of unit root tests has been predicated on the assumption that $\{x_t\}$ contains *at most* one unit root, i.e., that it is $I(1)$. If the null hypothesis of a unit root is not rejected, then it may be necessary to test whether the series contains a second unit root, i.e., whether it is $I(2)$ and thus needs differencing twice to induce stationarity. Unfortunately, the 'standard' testing procedure, on non-rejection of a unit root in the levels x_t, of testing whether the differences ∇x_t contain a unit root is not justified theoretically, as Dickey–Fuller type tests are based on the assumption of at most one unit root. If the true number of unit roots is greater than one, the empirical size of such tests is greater than the nominal size, so the probability of finding any, let alone all, unit roots is reduced.

Dickey and Pantula (1987) propose a sequence of tests that does have a theoretical justification when we assume that $\{x_t\}$ may contain more than one unit root. For example, suppose we assume that x_t contains a maximum of two unit roots. To test the null hypothesis of two unit roots against the alternative of one, we compare the t-ratio on β_2 from the regression

$$\nabla^2 x_t = \beta_0 + \beta_2 \nabla x_{t-1} + a_t,$$

with the τ_μ tables in Fuller (1976). If the null is rejected, we may then test the hypothesis of exactly one unit root against the alternative of none by comparing the t-ratio on β_1 from

$$\nabla^2 x_t = \beta_0 + \beta_1 x_{t-1} + \beta_2 \nabla x_{t-1} + a_t,$$

with the τ_μ distribution.

Example 3.4: Do UK interest rates contain two unit roots?

Figure 3.3 shows plots of the UK short and long interest rates from which the spread, analysed, for example, in example 3.1, was calculated. To test for the presence of at most two unit roots we first estimate, under the null hypothesis of exactly two unit roots, the regressions

$$\nabla^2 r_t = \underset{(0.098)}{0.079} - \underset{(0.082)}{0.983} \nabla r_{t-1},$$

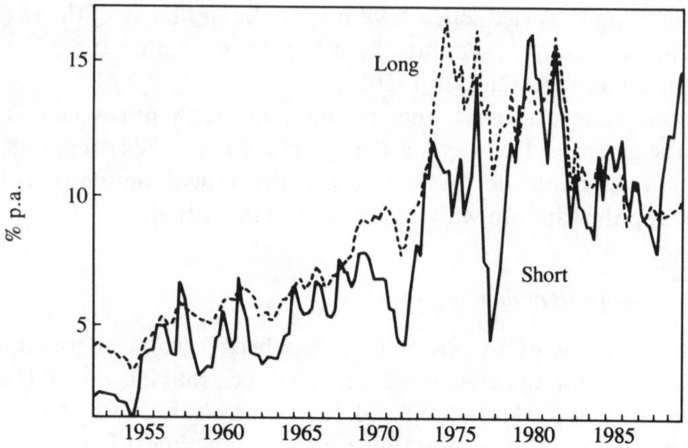

Figure 3.3 UK interest rates (quarterly 1952–88)

and

$$\nabla^2 R_t = \underset{(0.061)}{0.043} - \underset{(0.081)}{1.147} \nabla R_{t-1},$$

where r_t and R_t are the short and long rates respectively. The τ_μ statistics are computed to be, respectively, -11.97 and -14.10, thus conclusively rejecting the hypothesis of two unit roots. On estimating the regressions

$$\nabla^2 r_t = \underset{(0.226)}{0.469} - \underset{(0.028)}{0.052} r_{t-1} - \underset{(0.082)}{0.958} \nabla r_{t-1},$$

and

$$\nabla^2 R_t = \underset{(0.166)}{0.277} - \underset{(0.018)}{0.027} R_{t-1} - \underset{(0.081)}{1.137} \nabla R_{t-1},$$

however, we find that the τ_μ statistics from the estimates of β_1 are -1.90 and -1.52 respectively, thus providing no evidence against the hypothesis that both series contain a single unit root.

3.2 Decomposing time series: unobserved component models and signal extraction

3.2.1 Unobserved component models

If a time series is difference stationary, then it can be decomposed into a stochastic non-stationary, or trend, component and a stationary, or noise, component, i.e.

$$x_t = z_t + u_t. \tag{3.13}$$

Such a decomposition can be effected in many ways. For instance, Muth's (1960) classic example assumes that the trend component z_t is a random walk

$$z_t = \mu + z_{t-1} + v_t,$$

while u_t is white noise and independent of v_t, i.e., $u_t \sim WN(0,\sigma_u^2)$ and $v_t \sim WN(0,\sigma_v^2)$, with $E(u_t v_{t-i}) = 0$ for all i. It thus follows that ∇x_t is a stationary process

$$\nabla x_t = \mu + v_t + u_t - u_{t-1},$$

and has an ACF that cuts off at lag one with coefficient

$$\rho_1 = -\frac{\sigma_u^2}{\sigma_u^2 + 2\sigma_v^2}. \tag{3.14}$$

It is clear that $-0.5 \leq \rho_1 \leq 0$, the exact value depending on the relative sizes of these variances.

As a second example, consider the Poterba and Summers (1988) model for measuring mean reversion in stock prices. Rather than assume the noise component to be purely random, they allow it to follow an AR(1) process

$$u_t = \lambda u_{t-1} + a_t, \tag{3.15}$$

so that

$$\nabla x_t = \mu + v_t + (1 - \lambda B)^{-1}(1 - B)a_t.$$

∇x_t thus follows the ARMA(1,1) process

$$(1 - \lambda B)\nabla x_t = \theta_0 + (1 - \theta_1 B)e_t,$$

where $e_t \sim WN(0,\sigma_e^2)$, $\theta_0 = \mu(1-\lambda)$

$$\theta_1 = \{(1+\lambda^2)\sigma_v^2 + 2\sigma_a^2 - (1-\lambda)\sigma_v[4\sigma_a^2 + (1+\lambda)^2]^{\frac{1}{2}}\}/(2\sigma_a^2 + 2\lambda),$$

and

$$\sigma_e^2 = (\lambda\sigma_v^2 + \sigma_a^2)/\theta_1.$$

Muth's model is, obviously, a special case of this with $\lambda = 0$, and hence is characterised by ∇x_t following an MA(1) process with parameters

$$\theta_1 = \{2 + \kappa - (\kappa^2 + 4\kappa)^{\frac{1}{2}}\}/2,$$

where $\kappa = \sigma_v^2/\sigma_u^2$ is the *signal-to-noise* variance ratio, and $\sigma_e^2 = \sigma_u^2/\theta_1$.

Models of the form (3.13) are known as *unobserved component* (UC) models, a more general formulation for the components being

$$\nabla z_t = \mu + \gamma(B)v_t,$$

and (3.16)

$$u_t = \lambda(B)a_t,$$

where v_t and a_t are independent white-noise sequences with finite variances σ_v^2 and σ_a^2 and where $\gamma(B)$ and $\lambda(B)$ are stationary polynomials having no common roots. It can be shown that x_t will then have the form

$$\nabla x_t = \mu + \theta(B)e_t, \tag{3.17}$$

where $\theta(B)$ and σ_e^2 can be obtained from

$$\sigma_e^2 \frac{\theta(B)\theta(B^{-1})}{(1-B)(1-B^{-1})} = \sigma_v^2 \frac{\gamma(B)\gamma(B^{-1})}{(1-B)(1-B^{-1})} + \sigma_a^2 \lambda(B)\lambda(B^{-1}). \tag{3.18}$$

From this we see that it is not necessarily the case that the parameters of the components can be identified from knowledge of the parameters of (3.17) alone: in general, the components will not be identified. However, if z_t is restricted to be a random walk ($\gamma(B)=1$), the parameters of the UC model will be identified. This is clearly the case for Muth's model since σ_u^2 can be estimated by the lag one autocovariance of ∇x_t (the numerator of (3.14)) and σ_v^2 from the variance of ∇x_t (the denominator of (3.14)) and the estimated value of σ_u^2.

This example illustrates, however, that even though the variances are identified, such a decomposition may not always be feasible, for it is unable to account for positive first-order autocorrelation in ∇x_t. To do so requires relaxing either the assumption that z_t is a random walk, so that the trend component can contain both permanent and transitory movements, or the assumption that v_t and u_t are independent. If either of these assumptions is relaxed, the parameters of the Muth model will not be identified.

A trend component following a random walk is not as restrictive as it may first seem. Consider the Wold decomposition for ∇x_t

$$\nabla x_t = \mu + \psi(B)e_t = \mu + \sum_{j=0}^{\infty} \psi_j e_{t-j}. \tag{3.19}$$

Beveridge and Nelson (1981) show that x_t can be decomposed as (3.13) with

$$\nabla z_t = \mu + \left(\sum_{j=0}^{\infty} \psi_j\right) e_t,$$

and

$$-u_t = \left(\sum_{j=1}^{\infty} \psi_j\right) e_t + \left(\sum_{j=2}^{\infty} \psi_j\right) e_{t-1} + \left(\sum_{j=3}^{\infty} \psi_j\right) e_{t-2} + \ldots$$

Since e_t is white noise, the trend component is therefore a random walk with rate of drift equal to μ and an innovation equal to $(\sum_0^{\infty} \psi_j)e_t$, which is thus proportional to that of the original series. The noise component is clearly stationary, but since it is driven by the same innovation as the trend component, z_t and u_t must be *perfectly correlated*, in direct contrast to the Muth decomposition that assumes that they are independent.

In a more general context, it is possible for an x_t with Wold decomposition given by (3.19) to be written as (3.13) with z_t being a random walk and u_t being stationary and where the innovations to the two components are correlated to an arbitrary degree. However, only the Beveridge–Nelson decomposition is *guaranteed* to exist.

3.2.2 Signal extraction

Given a UC model of the form (3.13) and models for z_t and u_t, it is often useful to provide estimates of these two unobserved components: this is known as *signal extraction*. A MMSE estimate of z_t is an estimate $\hat{z}_t$ which minimises $E(\zeta_t^2)$, where ζ_t is the estimation error $z_t - \hat{z}_t$. From, for example, Pierce (1979), given the *infinite sample* $\{x_t\}_{-\infty}^{\infty}$, such an estimator is

$$\hat{z}_t = \nu_z(B)x_t = \sum_{j=-\infty}^{\infty} \nu_{zj} x_{t-j},$$

where the filter $\nu_z(B)$ is defined as

$$\nu_z(B) = \frac{\sigma_v^2 \gamma(B)\gamma(B^{-1})}{\sigma_e^2 \theta(B)\theta(B^{-1})},$$

in which case the noise component can be estimated as

$$\hat{u}_t = x_t - \hat{z}_t = [1 - \nu_z(B)]x_t = \nu_u(B)x_t.$$

For example, for the Muth model of a random walk overlaid with stationary noise, we have

$$\nu_z(B) = \frac{\sigma_v^2}{\sigma_e^2}(1-\theta B)^{-1}(1-\theta B^{-1})^{-1} = \frac{\sigma_v^2}{\sigma_e^2}\frac{1}{(1-\theta^2)}\sum_{j=-\infty}^{\infty} \theta^{|j|} B^j,$$

so that, using $\sigma_v^2 = (1-\theta)^2\sigma_e^2$, obtained by applying (3.18), we have

$$\hat{z}_t = \frac{(1-\theta)^2}{1-\theta^2}\sum_{j=-\infty}^{\infty} \theta^{|j|} x_{t-j}.$$

Thus, for values of θ close to unity, $\hat{z}_t$ will be given by a very long moving average of future and past values of x. If θ is close to zero, however, $\hat{z}_t$ will be almost equal to the most recently observed value x_t. From (3.14) it is clear that large values of θ correspond to small values of the signal–to–noise ratio, σ_v^2/σ_u^2: when the noise component dominates, a long moving average of x values provides the best estimate of trend, while if the noise component is only small the trend is given by the current position of x.

The estimation error, $\zeta_t = z_t - \hat{z}_t$, can be written as

$$\zeta_t = v_z(B)z_t - v_u(B)u_t,$$

and Pierce (1979) shows that ζ_t will be stationary if z_t and u_t are generated by processes of the form (3.16). In fact, ζ_t will follow the process

$$\zeta_t = \theta_\zeta(B)\xi_t,$$

where

$$\theta_\zeta(B) = \frac{\gamma(B)\lambda(B)}{\theta(B)},$$

$$\sigma^2_{\xi_t} = \frac{\sigma^2_a \sigma^2_v}{\sigma^2_e}$$

and $\xi_t \sim WN(0, \sigma^2_{\xi_t})$.

For the Muth model we thus have that ζ_t follows the AR(1) process

$$(1-\theta B)\zeta_t = \xi_t,$$

and the MSE of the optimal signal extraction procedure is

$$E(\zeta^2_t) = \frac{\sigma^2_a \sigma^2_v}{\sigma^2_e(1-\theta^2)}.$$

As noted earlier, if we are given only $\{x_t\}$ and its model, i.e., (3.17), then the models for z_t and u_t are in general unidentified. If x_t follows the IMA(1) process

$$(1-B)x_t = (1-\theta B)e_t, \tag{3.20}$$

then the most general signal–plus–*white*-noise UC model has z_t given by

$$(1-B)z_t = (1-\Theta B)v_t, \tag{3.21}$$

and for any Θ value in the interval $-1 \leq \Theta \leq \theta$ there exists values of σ^2_a and σ^2_v such that $z_t + u_t$ yields (3.20). It can be shown that setting $\Theta = -1$ minimises the variance of both z_t and u_t and is known as the *canonical decomposition* of x_t. Choosing this value implies that $\gamma(B) = 1 + B$, and we thus have

$$\hat{z}_t = \frac{\sigma^2_v(1+B)(1+B^{-1})}{\sigma^2_e(1-\theta B)(1-\theta B^{-1})},$$

and

$$(1-\theta B)\zeta_t = (1+B)\xi_t.$$

In this development we have assumed that in estimating z_t the future as well as the past of $\{x_t\}$ is available. In many situations it is necessary to estimate z_t given only data on x_s up to $s=t-m$, for finite m. This includes the problems of signal extraction based either on current data ($m=0$) or on recent data ($m<0$), and the problem of forecasting the signal ($m>0$). We thus need to extend the analysis to consider signal extraction given only the *semi-infinite* sample $\{x_s, s \leq t-m\}$. Pierce (1979) shows that, in this case, an estimate of z_t is given by

$$\hat{z}_t^{(m)} = \nu_z^{(m)}(B)x_t,$$

where

$$\nu_z^{(m)}(B) = \frac{(1-B)}{\sigma_e^2\theta(B)}\left[\frac{\sigma_v^2\gamma(B)\gamma(B^{-1})}{(1-B)\theta(B^{-1})}\right]_m,$$

in which we use the notation

$$[h(B)]_m = \sum_{j=m}^{\infty} h_j B^j.$$

Thus for the Muth model we have

$$\nu_z^{(m)}(B) = \frac{\sigma_v^2(1-B)}{\sigma_e^2(1-\theta B)}\left[\frac{(1-B)^{-1}}{(1-\theta B^{-1})}\right]_m,$$

and Pierce (1979) shows that this becomes, for $m \geq 0$

$$\nu_z^{(m)}(B) = \frac{\sigma_v^2 B^m}{\sigma_e^2(1-\theta)}\sum_{j=0}^{\infty}(\theta B)^j,$$

while for $m<0$

$$\nu_z^{(m)}(B) = \theta^{-m}\frac{\sigma_v^2 B^m}{\sigma_e^2(1-\theta)}\sum_{j=0}^{\infty}(\theta B)^j + \frac{1}{1-\theta B}\sum_{j=0}^{-m-1}\theta^j B^{-j}.$$

Thus, when either estimating z_t for the current time period ($m=0$) or forecasting z_t ($m>0$), we apply an exponentially weighted moving average to the observed series, beginning with the most recent data available, but not otherwise depending on the value of m. For $m<0$, when we are estimating z_t based on some, but not all, of the relevant future observations of x_t , the filter comprises two parts: the same filter as in the $m \geq 0$ case applied to the furthest forward observation but with a declining weight (θ^{-m}) placed upon it, and a second term capturing the additional influence of the observed future observations.

UC models can also be analysed within a *state space* framework, in which the Kalman filter plays a key role in providing both optimal forecasts and a method of estimating the unknown model parameters. In this framework, models such as the random walk–plus–white noise are known as *structural models*, and a thorough discussion of the methodological and technical ideas underlying such formulations is contained in the monograph by Harvey (1989), with more recent technical developments and applications being surveyed in Harvey and Shephard (1992).

Example 3.5: Estimating expected real rates of interest

An important example of the unobserved random walk buried in white noise is provided by the analysis of expected real rates of interest under the assumption of rational expectations or, equivalently, financial market efficiency: see, for example, Fama (1975) and Nelson and Schwert (1977). In this model, the unobservable expected real rate, z_t, is assumed to follow a driftless random walk, i.e., equation (3.21) with $\Theta=0$, and it differs from the observed real rate, x_t, by the amount of unexpected inflation, u_t, which, under the assumption of market efficiency, will be a white-noise process. The observed real rate will thus follow the ARIMA(0,1,1) process shown in (3.20).

Mills and Stephenson (1985) employ such a model to analyse the UK Treasury bill market over the period 1952 to 1982. They find that observed real Treasury bill rates are adequately modelled by the process

$$\nabla x_t=(1-0.678B)e_t, \quad \hat{\sigma}_e^2=14.5.$$

From the relationships linking σ_v^2 and σ_u^2 to θ and σ_e^2, it follows that the unobserved variances may be estimated as

$$\hat{\sigma}_v^2=(1-0.678)^2\hat{\sigma}_e^2=1.51$$
$$\hat{\sigma}_u^2=0.678\hat{\sigma}_e^2=9.85,$$

yielding a signal–to–noise variance ratio of $\kappa=\hat{\sigma}_v^2/\hat{\sigma}_u^2=0.15$, so that variations in the expected real rate are small compared to variations in unexpected inflation. Expected real rates based on information up to and including time t, i.e., $m=0$, can then be estimated using the exponentially weighted moving average

$$\hat{z}_t=\nu_z^{(0)}(B)x_t,$$

where

$$\nu_z^{(0)}(B)=\frac{\sigma_v^2}{\sigma_e^2(1-\theta)}\sum_{j=0}^{\infty}(\theta B)^j,$$

$$=0.476\sum_{j=0}^{\infty}(0.322B)^j.$$

Unexpected inflation can then be obtained as $\hat{u}_t=x_t-\hat{z}_t$. Figure 3.4 provides plots of x_t, $\hat{z}_t$ and $\hat{u}_t$, showing that the expected real rate is considerably smoother than the observed real rate, as was suggested by the

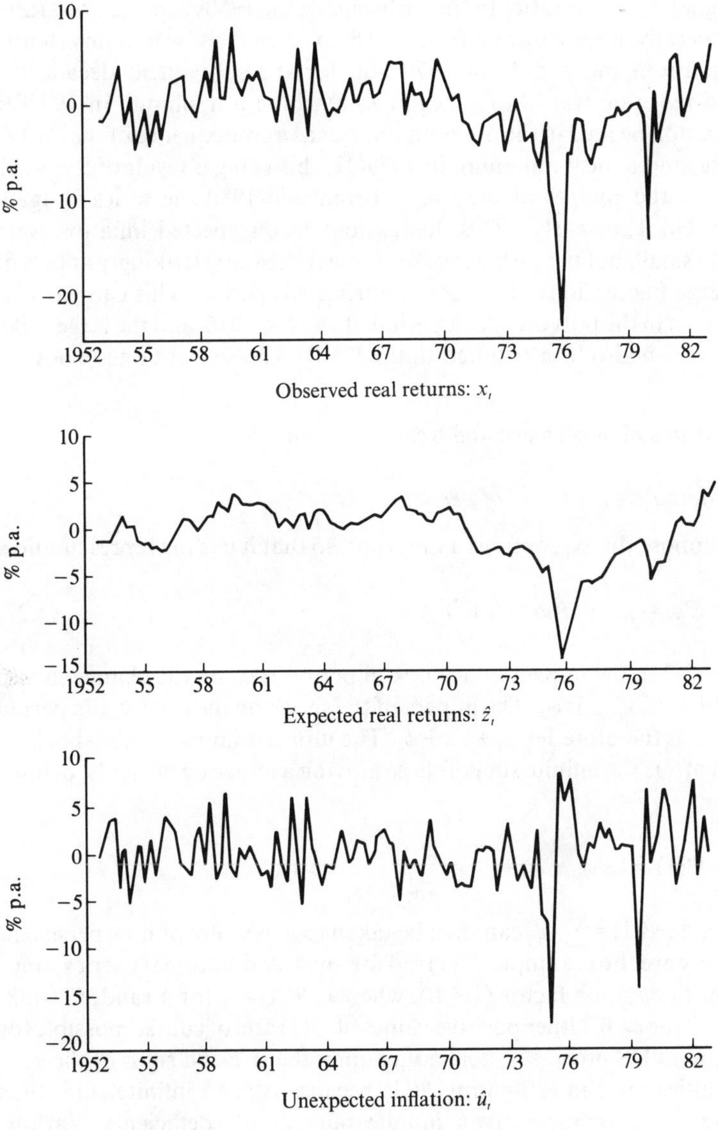

Figure 3.4 Treasury bill real return decomposition

small signal–to–noise ratio. In the early part of the 1950s expected real rates were generally negative, but from 1956 to 1970 they were consistently positive. From the middle of 1970 and for the subsequent decade the expected real rate was always negative, reaching a minimum in 1975Q1 after inflation peaked in the previous quarter as a consequence of the OPEC price rise, and a local minimum in 1979Q2, this being a result of the VAT increase in the Budget of that year. From mid-1980 the series is again positive. Until the early 1970s fluctuations in unexpected inflation were relatively small, but the turbulence of the next decade is strikingly apparent in the large fluctuations of the series during this period. This can particularly be seen in the post oil-shock period of 1974 to 1976, and the large spike in 1979Q2 can also be attributed to the VAT increase mentioned above.

3.3 Measures of persistence and trend reversion

3.3.1 Alternative measures of persistence

Let us suppose that x_t contains a unit root, so that it has the representation

$$\nabla x_t = \mu + \psi(B)a_t = \mu + \sum_{j=0}^{\infty} \psi_j a_{t-j}. \tag{3.22}$$

From (3.22), the impact of a shock in period t, a_t, on the change in x in period $t+k$, ∇x_{t+k}, is ψ_k. The impact of the shock on the *level* of x in period $t+k$, x_{t+k}, is therefore $1+\psi_1+\ldots+\psi_k$. The ultimate impact of the shock on the level of x is the infinite sum of these moving average coefficients, defined as

$$\Psi(1) = 1 + \psi_1 + \psi_2 + \ldots = \sum_{j=0}^{\infty} \psi_j.$$

The value $\Psi(1) = \sum_0^\infty \psi_j$ can then be taken as a measure of how persistent shocks to x are. For example, $\Psi(1)=0$ for any trend stationary series, since $\Psi(B)$ must contain a factor $(1-B)$, whereas $\Psi(1)=1$ for a random walk, since $\psi_j = 0$ for $j>0$. Other positive values of $\Psi(1)$ are, of course, possible for more general DS processes, depending upon the size and signs of the ψ_j.

Difficulties arise in estimating $\Psi(1)$ because it is an infinite sum, thus requiring the estimation of an infinite number of coefficients. Various measures have thus been proposed in the literature to circumvent this problem, two of the most popular being the *impulse response* measure proposed by Campbell and Mankiw (1987) and the *variance ratio* of Cochrane (1988).

Campbell and Mankiw offer a measure of $\Psi(1)$ based on approximating $\Psi(B)$ by a ratio of finite order polynomials. This is possible because, since it is assumed that ∇x_t is a linear stationary process, it follows that it has an ARMA(p,q) representation

$$\phi(B)\nabla x_t = \theta_0 + \theta(B)a_t, \tag{3.23}$$

where $\phi(B) = 1 - \phi_1 B - \ldots - \phi_p B^p$ and $\theta(B) = 1 - \theta_1 B - \ldots - \theta_q B^q$. Equation (3.22) is then interpreted as the moving average representation, or *impulse response function*, of ∇x_t

$$\nabla x_t = \phi(1)^{-1}\theta_0 + \phi(B)^{-1}\theta(B)a_t.$$

From the equality $\Psi(B) = \phi(B)^{-1}\theta(B)$, the measure $\Psi(1)$ can then be calculated directly as $\Psi(1) = \theta(1)/\phi(1)$.

Cochrane (1988), on the other hand, proposes a non-parametric measure of persistence known as the *variance ratio*, defined as $V_k = \sigma_k^2/\sigma_1^2$, where

$$\sigma_k^2 = k^{-1}V(x_t - x_{t-k}) = k^{-1}V(\nabla_k x_t),$$

$\nabla_k = 1 - B^k$ being the kth differencing operator. This measure is based on the following argument. If x_t is a pure random walk, then the variance of its kth differences will grow linearly with k: using the fact that $\nabla_k = \nabla(1 + B + \ldots + B^{k-1})$

$$\begin{aligned} V(\nabla_k x_t) &= V((x_t - x_{t-1}) + (x_{t-1} - x_{t-2}) + \ldots + (x_{t-k+1} - x_{t-k})) \\ &= \sum_{j=1}^{k} V(x_{t-j+1} - x_{t-j}) = k\sigma^2. \end{aligned}$$

If, on the other hand, x_t is trend stationary, the variance of its kth-differences approaches a constant, this being twice the unconditional variance of the series

$$\begin{aligned} V(\nabla_k x_t) &= V\left(k\mu + \sum_{j=0}^{\infty}\psi_j a_{t-j} - \sum_{j=0}^{\infty}\psi_j a_{t-j-k}\right) \\ &= V\left(\sum_{j=0}^{k-1}\psi_j a_{t-j} + \sum_{j=0}^{\infty}(\psi_{k+j} - \psi_j)a_{t-j-k}\right) \\ &= \sigma^2\sum_{j=0}^{k-1}\psi_j + \sigma^2\sum_{j=0}^{\infty}(\psi_{k+j} - \psi_j), \end{aligned}$$

which, as $k \to \infty$

$$\to 2\sigma^2\sum_{j=0}^{\infty}\psi_j = 2V(x_t).$$

Cochrane thus suggests plotting a sample estimate of σ_k^2 as a function of k. If x_t is a random walk, the plot should be constant at σ^2, whereas if x_t is trend stationary the plot should decline towards zero. If fluctuations in x_t

are partly permanent and partly temporary, so that the series can be modelled as a combination of random walk and stationary components, the plot of σ_k^2 versus k should settle down to the variance of the innovation to the random walk component.

In providing a sample estimate of σ_k^2, Cochrane corrects for two possible sources of small-sample bias. First, the sample mean of ∇x_t is used to estimate the drift term μ at all k, rather than a different drift term at each k being estimated from the mean of the k-differences. Second, a degrees of freedom correction $T/(T-k-1)$ is included, for, without this, σ_k^2 will decline towards zero as $k \to T$ because a variance cannot be taken with one observation. These corrections produce an estimator of σ_k^2 that is unbiased when applied to a pure random walk with drift. The actual formula used to compute the estimator from the sample $\{x_t\}_0^T$ is (Cochrane, 1988, equation (A3), p. 917)

$$\hat{\sigma}_k^2 = \frac{T}{k(T-k)(T-k+1)} \sum_{j=k}^{T} \left[x_j - x_{j-k} - \frac{k}{T}(x_T - x_0) \right]^2 .$$

From Cochrane (1988), the asymptotic standard error of $\hat{\sigma}_k^2$ is $(4k/3T)^{.5}\hat{\sigma}_k^2$. The variance ratio can then be estimated as $\hat{V}_k = \hat{\sigma}_k^2/\hat{\sigma}_1^2$.

Cochrane shows that V_k can also be written as

$$V_k = 1 + 2\sum_{j=1}^{k-1} \frac{k-j}{k} \rho_j ,$$

and that, since

$$\lim_{k\to\infty} \sigma_k^2 = \sigma_{\nabla x}^2 = |\Psi(1)|^2 \sigma^2 = (\textstyle\sum \psi_j)^2 \sigma^2 ,$$

the *limiting variance ratio*, V, can be defined as

$$V \equiv \lim_{k\to\infty} V_k = 1 + 2\sum_{j=1}^{\infty} \rho_j .$$

While an estimate of $\Psi(1)$ cannot be obtained directly from an estimate of σ_k^2, Cochrane's measure does provide a lower bound on $\Psi(1)$. To see this, note that V can be written as

$$V = (\sigma^2/\sigma_1^2)\,|\Psi(1)|^2 .$$

By defining $R^2 = 1 - (\sigma^2/\sigma_1^2)$, the fraction of the variance that is predictable from knowledge of the past history of ∇x_t, we have

$$\Psi(1) = \sqrt{\frac{V}{1-R^2}} . \tag{3.24}$$

This shows that the square root of V is a lower bound on $\Psi(1)$: the more highly predictable is ∇x_t, the greater the difference between the two measures. Campbell and Mankiw (1987) suggest computing an approximate estimate of $\Psi(1)$ by replacing R^2 in (3.24) with the square of the first sample autocorrelation of ∇x_t, r_1^2. Since r_1^2 is an underestimate of R^2, this estimate will tend to understate $\Psi(1)$.

The variance ratio can also be used to obtain an estimate of $\Psi(1)$ from an unobserved components model. Campbell and Mankiw (1987) note that the assumption of independence between the trend and noise components implies that V can be written as a weighted average of the variance ratios of the two components. If these are denoted V_z and V_u respectively, then

$$V = \delta V_z + (1-\delta)V_u,$$

where $\delta = \sigma_v^2/\sigma_1^2$. Since the UC model assumes that z_t is a random walk and u_t is stationary, $V_z = 1$ and $V_u = 0$, so that $V = \delta = \sigma_v^2/\sigma_1^2$. A measure of $\Psi(1)$ can then be obtained from equation (3.24) by replacing R^2 by an estimate of r_1^2 as before.

Each of these approaches to estimating persistence has its advantages and drawbacks and hence its proponents and critics. Impulse response functions have the advantage of using the ARMA modelling approach, which is well known and developed. Cochrane (1988), however, criticises the use of such models in this context because they are designed to capture *short-run* dynamics (recall their development for short-term forecasting by Box and Jenkins, 1976), rather than the *long-run* correlations that are of interest here.

The UC models have also been criticised by Cochrane on similar grounds, in that the identifying restrictions required to estimate long-run behaviour are themselves based on short-run dynamics. Furthermore, such models rule out highly persistent processes a priori, and Nelson (1988) has provided Monte Carlo evidence to suggest that they also have a tendency to indicate (incorrectly) that a series consists of cyclical variations around a smooth trend when the data is actually generated by a random walk, thus biasing estimates of $\Psi(1)$ downwards. Nonetheless, proponents of UC models argue that they do, in fact, capture long-run correlations much better than ARMA models and estimates of $\Psi(1)$ can easily be obtained using the approach suggested by Campbell and Mankiw (1987).

The non-parametric measure provides only an approximate estimate of $\Psi(1)$, is accompanied by large standard errors, and the 'window size' k can be difficult to determine. Nevertheless, Cochrane (1988) argues that it is the only measure which is explicitly designed to model long-run dynamics and to isolate random walk components without being contaminated by short-run correlations.

3.3.2 Testing for mean reversion

Whether a random walk is present in a financial time series has been shown to be a question of some importance. The unit root tests discussed in section 3.1 are, of course, one approach to testing whether a series contains a random walk component. As we have seen, however, such tests can have difficulties in detecting some important departures from a random walk, and the associated distributions of the test statistics have awkward dependencies on nuisance parameters.

When the null hypothesis under examination is that a series is generated by a random walk with independently and identically distributed (iid) normal increments, a test based on the variance ratio may be preferred. Consider again the observed series $\{x_t\}_0^T$ and suppose that x_t is generated by the random walk

$$x_t = \theta + x_{t-1} + a_t,$$

where $a_t \sim NID(0,\sigma^2)$. From the previous section, an unbiased estimator of σ^2 is given by

$$\hat{\sigma}_1^2 = \frac{1}{(T-1)} \sum_{t=1}^{T} \left[x_t - x_{t-1} - \frac{1}{T}(x_T - x_0) \right]^2 .$$

Now, under our null hypothesis of a random walk, the variance of the kth differences of x_t

$$\hat{\sigma}_k^2 = \frac{T}{k(T-k)(T-k+1)} \sum_{t=k}^{T} \left[x_t - x_{t-k} - \frac{k}{T}(x_T - x_0) \right]^2 ,$$

will also be an unbiased estimator of σ^2. Lo and MacKinlay (1988, 1989) define the test statistic

$$M(k) = \hat{\sigma}_k^2 / \hat{\sigma}_1^2 - 1 = \hat{V}_k - 1$$

and show that

$$z_1(k) = M(k) \cdot \left(\frac{2(2k-1)(k-1)}{3nk^2} \right)^{-\frac{1}{2}}, \tag{3.25}$$

where $n = T/k$, is asymptotically $N(0,1)$.

Lo and MacKinlay also derive a version of the variance ratio test that is robust to heteroskedasticity. In this case the test statistic becomes

$$z_2(k) = M(k) \cdot \Omega^{-\frac{1}{2}}(k),$$

where

$$\Omega(k) = \sum_{j=1}^{k-1} \left[\frac{2(k-j)}{k} \right]^2 \cdot \delta(j),$$

$$\delta_j = \frac{\sum_{t=j+1}^{T} a_{0t} a_{jt}}{\left(\sum_{t=1}^{T} a_{0t}\right)^2}, \quad a_{jt} = \left(x_{t-j} - x_{t-j-1} - \frac{1}{T}(x_T - x_0)\right)^2.$$

The δ_j are heteroskedastic-consistent estimators of the asymptotic variances of the estimated autocorrelations of ∇x_t.

Lo and MacKinlay (1989) find that this large-sample normal approximation works well when k is small and T is large. They emphasise, however, that it can be unsatisfactory for large k because the empirical distribution of $z_1(k)$ is highly skewed in these circumstances: although the empirical sizes of the test statistic are close to their nominal values, almost all the rejections occur in the upper tail of the distribution. It is thus clear that the normal approximation to the distribution of $z_1(k)$ is likely to be of only limited practical use. As a consequence, the empirical distributions of the test statistics need to be evaluated by simulation.

In any case, the asymptotic normality of $M(k)$ relies on fixing k and allowing T to increase, so that $k/T \to 0$. Richardson and Stock (1989), however, consider a different perspective in which k is allowed to tend asymptotically to a non-zero fraction (δ) of T, i.e., $k/T \to \delta$. Under this asymptotic theory, $M(k)$ has a limiting distribution that is not normal, but has a representation in terms of functionals of Brownian motion, which under the null does not depend on any unknown parameters. Richardson and Stock argue that the $k/T \to \delta$ theory provides a much better approximation to the finite sample distribution of $M(k)$ than does the fixed k theory.

Lo and MacKinlay (1989) examine the power of the variance ratio test against three plausible alternatives. The first is that x_t is generated by a trend stationary AR(1) process (as has been suggested by Shiller (1981a) as a model of stock market 'fads')

$$x_t = \theta + \phi x_{t-1} + \omega t + a_t, \quad 0 < \phi < 1 \tag{3.26}$$

As discussed in previous sections, an obvious test is the τ_τ statistic. A second test is suggested from noting that, as shown above, an alternative expression for $\hat{V}_k$ is

$$\hat{V}_k = 1 + 2\sum_{j=1}^{k-1} \frac{k-j}{k} r_j. \tag{3.27}$$

The Q^* statistic introduced in chapter 2 is of similar form, being in this case

$$Q^*(k-1) = T\sum_{j=1}^{k-1} r_{k-1}^2 \sim \chi_{k-1}^2,$$

and thus provides another contender for a test statistic. Lo and MacKinlay find that $\hat{V}_k$ is comparable in power to τ_τ, and both are considerably more powerful than $Q^*(k-1)$.

Two unit root alternatives are also considered by Lo and MacKinlay. The first is the Poterba and Summers (1988) UC model with the stationary component being the AR(1) process (3.15), the second is the ARIMA(1,1,0) process

$$\nabla x_t = \varphi \nabla x_{t-1} + a_t.$$

The variance ratio test is found to be considerably more powerful than the other two when the variance of the stationary component is larger than that of the random walk, i.e., when the signal-to-noise ratio is small. Not surprisingly, since this alternative obviously possesses a unit root, the variance ratio is found to be more powerful than the other two, except when k is large, in which case the Q^* test dominates.

Example 3.6: Testing for mean reversion in UK stock returns

Poterba and Summers (1988) investigate mean reversion in stock returns by noting that, if the logarithm of the stock price (including cumulated dividends), denoted x_t, follows a random walk, then the variance of returns should be proportional to the return horizon. With monthly data, the k-month return, R_t^k, is defined as

$$R_t^k = x_t - x_{t-k} = \sum_{i=0}^{k-1} R_{t-i},$$

where $R_t = x_t - x_{t-1}$ is the return in month t.

Table 3.1 presents $M(k)$ statistics for nominal returns calculated from the monthly UK *FTA All Share* index (see example 2.3). We note first that all $M(k)$ values are positive ($\hat{V}_k$ values are greater than unity), thus indicating the presence of mean *aversion*, rather than reversion. Use of the asymptotic normal distribution to compute $z_1(k)$ and $z_2(k)$ statistics finds that almost all of these values are not significantly different from zero. But this is precisely the case where assuming asymptotic normality of the test statistics is most likely to provide misleading inferences, for the empirical distribution of $M(k)$ is known to be highly skewed to the right.

We therefore also provide appropriate percentiles of the empirical distributions of $M(k)$ obtained by Monte Carlo simulation using the Richardson and Stock (1989) $k/T \rightarrow \delta$ asymptotic theory (Mills, 1991a, provides further details). We do indeed see that as k increases, so the distributions become skewed, and we are now able to reject the random walk null at low levels of significance for most values of k. It would therefore appear that not only are UK stock returns predictable, as was found in example 2.3, but also that they are positively correlated at long horizons, thus rejecting the mean reversion model of stock returns as well as the random walk.

Table 3.1 *Variance ratio test statistics for UK stock returns (monthly data 1965–90)*

k	M	$p_1(M)$	$p_2(M)$	90%	95%	97.5%	99%
12	0.41	0.026	0.081	0.18	0.25	0.30	0.40
24	0.38	0.109	0.171	0.23	0.34	0.46	0.60
36	0.37	0.169	0.218	0.21	0.37	0.52	0.74
48	0.36	0.215	0.251	0.21	0.39	0.56	0.77
60	0.45	0.187	0.221	0.12	0.34	0.53	0.81
72	0.49	0.187	0.215	0.08	0.30	0.53	0.79
84	0.44	0.230	0.251	0.02	0.26	0.50	0.81
96	0.37	0.281	0.298	−0.06	0.16	0.37	0.65

Note:
$p_1(\cdot)$ and $p_2(\cdot)$ denote the probability under the null hypothesis of observing a larger variance ratio than that observed, and are calculated using the asymptotic normal distributions $z_1(\cdot)$ and $z_2(\cdot)$. 90%, ..., 99% are percentiles of the empirical distributions of $M(k)$ computed under the $k/T \rightarrow \delta$ asymptotic theory using $NID(0,1)$ returns with 5000 replications for each k: see Mills (1991a) for details.

3.4 Fractional integration and long memory processes

3.4.1 ARFIMA models

Much of the analysis of financial time series considers the case when the order of differencing, d, is either 0 or 1. If the latter, x_t is $I(1)$ and its ACF declines linearly. If the former, x_t is stationary ($I(0)$) and its ACF will exhibit an exponential decay: observations separated by a long time span may, therefore, be assumed to be independent, or at least nearly so. As we have seen, $I(1)$ behaviour of financial time series is an implication of many models of efficient markets, and the previous sections have discussed the analysis of such behaviour in considerable detail.

However, many empirically observed time series, although satisfying the assumption of stationarity (perhaps after some differencing transformation), seem to exhibit a dependence between distant observations that, although small, is by no means negligible. Such series are particularly found in hydrology, where this 'persistence' is known as the Hurst phenomenon, (see, for example, Mandelbrot and Wallis, 1969, and Hosking, 1984), but also many economic time series exhibit similar characteristics of extremely long persistence. This may be characterised as a tendency for large values to be followed by large values of the same sign in such a way that the series seem to go through a succession of 'cycles', including long cycles whose length is comparable to the total sample size.

This viewpoint has been persuasively argued by Mandelbrot (1969, 1972) in extending his work on non-Gaussian (marginal) distributions in economics, particularly financial prices (Mandelbrot, 1963), to an exploration of the structure of serial dependence in economic time series. While Mandelbrot considered processes that were of the form of discrete time fractional Gaussian noise, attention has focused recently on an extension of the ARIMA class to model long-term persistence.

We have so far considered only integer values of d. If d is non-integer, however, x_t is said to be *fractionally integrated,* and models for such values of d are referred to as ARFIMA (AR Fractionally IMA) by Diebold and Rudebusch (1989). This notion of fractional integration seems to have been proposed independently by Hosking (1981) and Granger and Joyeux (1980). To make the concept operational, we may use the binomial series expansion

$$\begin{aligned}\nabla^d = (1-B)^d &= \sum_{k=0}^{\infty} \begin{bmatrix} d \\ k \end{bmatrix} (-B)^k \\ &= 1 - dB - \frac{d(1-d)}{2!}B^2 - \frac{d(1-d)(2-d)}{3!}B^3 - \ldots\end{aligned} \tag{3.28}$$

How does the ARFIMA model incorporate 'long memory' behaviour? For $0 < d < \frac{1}{2}$, it can be shown that its ACF declines hyperbolically to zero, i.e., at a much slower rate than the exponential decay of a standard ARMA ($d=0$) process. For $d \geq \frac{1}{2}$, the variance of x_t is infinite, and so the process is non-stationary. Examples of how autocorrelations vary with d are provided in Hosking (1981), Diebold and Rudebusch (1989) and Lo (1991). Typically, autocorrelations from ARFIMA processes remain noticeably positive at very high lags, long after the autocorrelations from $I(0)$ processes have declined to (almost) zero.

The intuition behind the concept of long memory and the limitation of the integer-d restriction emerge more clearly in the frequency domain. $\{x_t\}$ will display long memory if its spectral density, $f_x(\vartheta)$, increases without limit as the frequency ϑ tends to zero

$$\lim_{\vartheta \to 0} f_x(\vartheta) = \infty .$$

If $\{x_t\}$ is ARFIMA then $f_x(\vartheta)$ behaves like ϑ^{-2d} as $\vartheta \to 0$, so that d parameterises its low-frequency behaviour. When $d=1, f_x(\vartheta)$ thus behaves like ϑ^{-2} as $\vartheta \to 0$, whereas when the integer-d restriction is relaxed a much richer range of spectral behaviour near the origin becomes possible. Indeed, the 'typical spectral shape' of economic time series (Granger, 1966), which exhibits monotonically declining power as frequency increases (except at seasonals), is well captured by an $I(d)$ process with $0<d<1$. Moreover, although the levels of many series have spectra that appear to be infinite at the origin, and so might seem to warrant first differencing, after such differencing they often have no power at the origin. This suggests that first differencing takes out 'too much' and that using a fractional d is thus a more appropriate form of 'detrending'.

3.4.2 Estimation of ARFIMA models

Since it is the parameter d that enables long-term persistence to be modelled, the value chosen for it is obviously crucial in any empirical application. Typically, this value will be unknown and must therefore be estimated. A number of methods have been proposed, but there is yet to become a large enough body of experimental evidence to suggest which are the clearly superior techniques, although this will surely become available before long.

Early suggestions are summarised in Mills (1990, chapter 11.7), Pagan and Wickens (1989) and Sowell (1992a). A popular approach is that proposed by Geweke and Porter-Hudak (1983). By writing

$$(1-B)^d x_t = (1-B)(1-B)^{d-1} x_t = (1-B)^{\tilde{d}} z_t,$$

where $z_t = (1-B)x_t$ and $\tilde{d} = d-1$, the ARFIMA model can be written (ignoring the constant θ_0 for convenience) as

$$(1-B)^{\tilde{d}} z_t = \phi^{-1}(B)\theta(B)a_t = u_t.$$

The spectral density of z_t is given by

$$f_z(\vartheta) = |1-\exp(-i\vartheta)|^{-2\tilde{d}} f_u(\vartheta) = [4\sin^2(\vartheta/2)]^{-\tilde{d}} f_u(\vartheta),$$

where $f_u(\vartheta)$ is the spectral density of the stationary process u_t. It then follows that

$$\ln\{f_z(\vartheta)\} = \ln\{f_u(\vartheta)\} - \tilde{d}\ln\{4\sin^2(\vartheta/2)\}$$

and, given the sample $\{z_t\}_1^T$, this leads Geweke and Porter-Hudak to propose estimating $\tilde{d}$ by regressing the periodogram $I_T(\vartheta_j)$ at frequencies $\vartheta_j = 2\pi j/T$, where $0<k_1 \leq j \leq K << T$, against a constant and $\ln\{4\sin^2(\vartheta_j/2)\}$. As pointed out by Pagan and Wickens (1989), Kunsch (1986) shows that frequencies around the origin need to be excluded to get a consistent estimator, and also shows how k_1 should vary with T. K should also expand with sample size and setting $K = T^{0.5}$ has been found to work well.

Having obtained an estimate of $\tilde{d}$, and hence d, x_t can be transformed by the long-memory filter (3.28), truncated at each point to the available sample. The transformed series is then modelled as an ARMA process. Further details of this procedure and discussion of its properties may be found in Geweke and Porter-Hudak (1983) and Diebold and Rudebusch (1989).

Sowell (1992a, 1992b) discusses joint maximum likelihood estimation of d and the ARMA parameters and presents Monte Carlo experiments which show that, when the correct model specification is known, maximum likelihood gives more accurate estimates than Geweke and Porter-Hudak's method. However, Sowell emphasises that, when the specification is uncertain, as it usually will be in practice, it remains an open question as to which of the methods is superior.

Example 3.7: Is there evidence of fractional integration in exchange and interest rates?

In example 3.1 we found that the weekly dollar/sterling exchange rate contained a unit root, while in example 3.4 a similar finding was obtained for the UK long interest rate. Each series was therefore subjected to the Geweke and Porter-Hudak (1983) technique for estimating d. On setting $k_1 = 3$, estimates of d were obtained for a wide range of K values: figure 3.5 provides plots of $\hat{d}$, with two standard error bounds, over the interval $15 \leq K \leq 40$ for both series. For the exchange rate, the values of $\hat{d}$ quickly settle down to just above unity, thus providing no evidence against the unit root hypothesis. For the long interest rate, however, the $\hat{d}$ values appear to settle down to around 0.85, with $d = 1$ only just being included in the two standard error bounds.

ARFIMA models for this series were then obtained and compared for the values $d = 1$ and $d = 0.85$. For $d = 1$, there was marginal evidence that the series is generated by an ARIMA(0,1,1) process

$$\nabla x_t = a_t - \underset{(0.082)}{0.141} a_{t-1},$$

whereas, for $d = 0.85$, fractionally integrated white noise is obtained, since the analogous ARFIMA(0,0.85,1) process was estimated to be

$$\nabla^{0.85} x_t = a_t + \underset{(0.082)}{0.027} a_{t-1}.$$

3.4.3 Robust tests of long-term memory

An alternative approach to detecting the presence of long-term memory in a time series, or long-range dependence as it is often known, is to use the

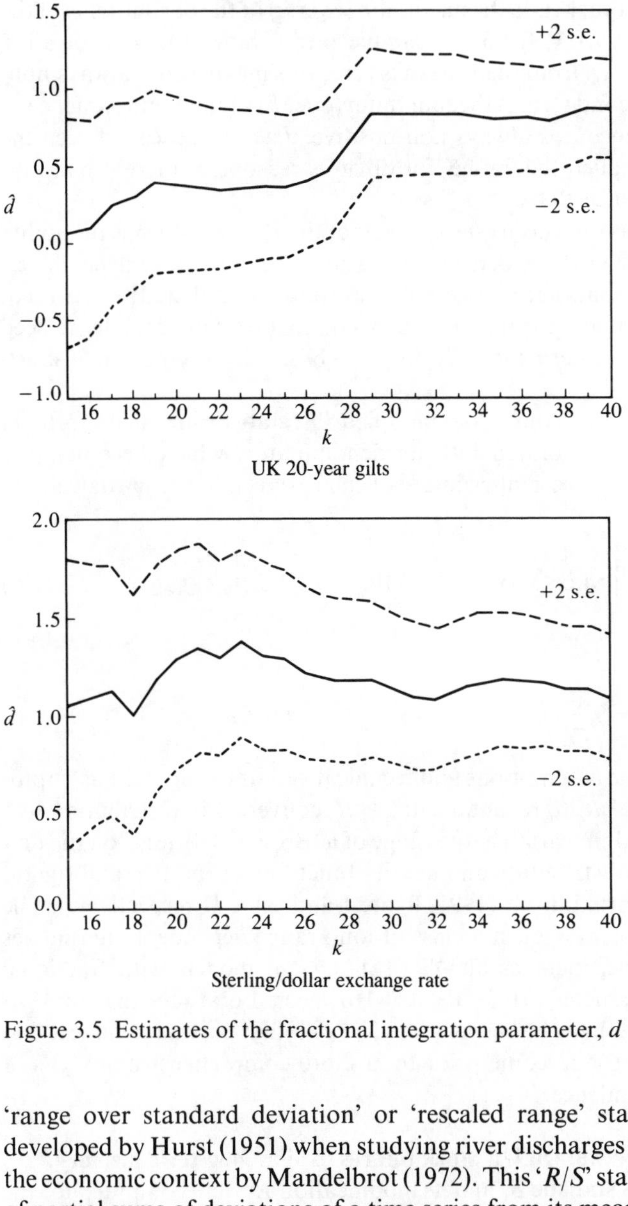

Figure 3.5 Estimates of the fractional integration parameter, d

'range over standard deviation' or 'rescaled range' statistic, originally developed by Hurst (1951) when studying river discharges and proposed in the economic context by Mandelbrot (1972). This 'R/S' statistic is the range of partial sums of deviations of a time series from its mean, rescaled by its standard deviation, i.e.

$$R_0 = s^{-1}\left[\underset{1\leq k\leq T}{\text{Max}}\sum_{t=1}^{k}(x_t-\bar{x}) - \underset{1\leq k\leq T}{\text{Min}}\sum_{t=1}^{k}(x_t-\bar{x})\right]. \tag{3.29}$$

The first term in brackets is the maximum (over k) of the partial sums of the first k deviations of x_t from the sample mean. Since the sum of all T deviations of the x_ts from their mean is zero, this maximum is always non-negative. The second term is the minimum (over k) of the same sequence of partial sums: hence it is always non-positive. The difference between the two quantities, called the 'range' for obvious reasons, is therefore always non-negative: hence $R_0 \geq 0$.

Although it has long been established that the R/S statistic has the ability to detect long-range dependence, it is sensitive to short-range dependence, so that any incompatibility between the data and the predicted behaviour of the R/S statistic under the null hypothesis of no long-run dependence need not come from long-term memory, but may be merely a symptom of short-term autocorrelation.

Lo (1991) therefore considers a modified R/S statistic in which short-run dependence is incorporated into its denominator, which becomes (the square root of) a consistent estimate of the variance of the partial sum in (3.29)

$$R_q = \hat{\sigma}_q^{-1}\left[\underset{1\leq k\leq T}{\text{Max}} \sum_{t=1}^{k}(x_t-\bar{x}) - \underset{1\leq k\leq T}{\text{Min}} \sum_{t=1}^{k}(x_t-\bar{x})\right], \qquad (3.30)$$

where

$$\hat{\sigma}_q^2 = s^2 + 2T\sum_{j=1}^{q} \omega_{qj} r_j, \quad \omega_{qj} = 1 - \frac{j}{q+1}, \quad q<T.$$

Lo provides the assumptions and technical details to allow the asymptotic distribution of R_q to be obtained: $T^{-\frac{1}{2}}R_q$ converges in distribution to a well-defined random variable (the range of a 'Brownian bridge' on the unit interval), whose distribution and density functions are plotted and significance levels reported in Lo (1991, figure I and table II respectively). The statistic is consistent against a class of long-range dependent alternatives that, for example, includes all ARFIMA(p,d,q) models with fractional differencing parameter $-0.5 \leq d \leq 0.5$. However, Lo argues that the R/S approach may perhaps be best regarded as a kind of portmanteau test that may complement, and come prior to, a more comprehensive analysis of long-range dependence.

Example 3.8: Do UK stock returns exhibit long-term memory?

The R/S statistic R_0 and its modification R_q were computed for the *FTA All Share* nominal returns series originally analysed in example 2.3 and which was found to be mean averting in example 3.6. $T^{-\frac{1}{2}}R_0$ was calculated to be 2.073 and, since a 95 per cent confidence interval for this statistic is, from Lo (1991, table II), (0.809, 1.862), we can reject the hypothesis that returns are iid in favour of the alternative that they exhibit long-range dependence.

However, from our previous examples, we know that returns are not iid: an AR(3) process provides the best-fitting model within the ARMA class, while the iid null was also rejected against a long-horizon alternative. We thus calculate the modified statistics R_3 and R_{12}, their normalised values being 1.876 and 1.713. There is now less evidence to reject the null hypothesis that returns are short-range dependent and this non-rejection is confirmed by using the Geweke and Porter-Hudak method of estimating the fractional differencing parameter d assuming an ARFIMA representation for returns, which is, of course, a particular class of long-memory model: using the same range of frequencies as in example 3.7, the estimated value of d was never significantly different from zero.

4 Univariate non-linear stochastic models

As we have seen in previous chapters, financial time series often appear to be well approximated by random walks. The relationship between random walks and the theory of efficient capital markets was briefly discussed in chapter 1, where it was argued that the random walk assumption that asset price changes are independent is usually too restrictive to be consistent with a reasonably broad class of optimising models: what is required is that a variable related to the asset price is a martingale.

Martingales and random walks are discussed formally in section 1, with tests of the random walk hypothesis being the subject of section 2. The relaxation of the assumption that changes in a time series must be independent and identically distributed allows the possibility of examining non-linear stochastic processes and the remainder of the chapter thus introduces various non-linear models that are becoming increasingly used in the analysis of financial time series. Stochastic variance models are discussed in section 3, ARCH processes in section 4, and bilinear and other related non-linear models in sections 5 and 6. Section 7 then considers chaotic models while, finally, section 8 looks at some tests for non-linearity.

4.1 Martingales, random walks and non-linearity

A *martingale* is a stochastic process that is a mathematical model of a 'fair game'. The term martingale, which also denotes part of a horse's harness or a ship's rigging, refers in addition to a gambling system in which every losing bet is doubled, a usage that may be felt to be rather apposite when considering the behaviour of financial data!

A martingale may be formally defined as a stochastic process $\{x_t\}_0^t$ having the following properties:

(a) $E[|x_t|]<\infty$ for each t;

(b) $E[x_t|\mathscr{H}_s]=x_s$, whenever $s\leq t$, where $\mathscr{H}_s$ is the σ-algebra comprising events determined by observations over the interval $[0,t]$, so that $\mathscr{H}_s\subseteq\mathscr{H}_t$ when $s\leq t$. This is known as the 'martingale property'.

While the 'history' $\{\mathscr{H}_t\}_0^t$ can, in general, include observations on any number of variables, it is often restricted to be just the past history of $\{x_t\}_0^t$ itself, i.e., $\mathscr{H}_t = \sigma(x_s;\ s \leq t)$.

Written as

$$E[x_t - x_s | \mathscr{H}_s] = 0, \quad s \leq t, \tag{4.1}$$

the martingale property implies that the MMSE forecast of a future increment of a martingale is zero. This property can be generalised to situations, quite common in finance, where

$$E[x_t - x_s | \mathscr{H}_s] \geq 0, \quad s \leq t,$$

in which case we have a *submartingale*, and to the case where the above inequality is reversed, giving us a *supermartingale*.

The martingale given by (4.1) can be written equivalently as

$$x_t = x_{t-1} + a_t,$$

where a_t is the martingale increment or *martingale difference*. When written in this form, the sequence $\{x_t\}_0^t$ looks superficially identical to the random walk, a model that was first introduced formally in chapter 2. There a_t was defined to be a stationary and uncorrelated sequence drawn from a fixed distribution, i.e., to be white noise. As was discussed in chapter 2.4, alternative definitions are, however, possible: a_t could be defined to be strict white noise, so that it is both a stationary and independent sequence, rather than just being uncorrelated. Moreover, it is possible for a_t to be uncorrelated but not necessarily stationary. While the white-noise assumptions rule this out, such behaviour is allowed for martingale differences: this implies that there could be dependence between higher conditional moments, most notably conditional variances.

The possibility of this form of dependence in financial time series, which often go through protracted quiet periods interspersed with bursts of turbulence, leads naturally to the consideration of *non-linear* stochastic processes capable of modelling such volatility. Non-linearity can, however, be introduced in many other ways, some of which may violate the martingale model. As an illustration, suppose that x_t is generated by the process $\nabla x_t = \eta_t$, with η_t being defined as

$$\eta_t = a_t + \beta a_{t-1} a_{t-2},$$

where a_t is strict white noise. It follows immediately that η_t has zero mean, constant variance, and ACF given by

$$E(\eta_t \eta_{t-k}) = E(a_t a_{t-k} + \beta a_{t-1} a_{t-2} a_{t-k} + \beta a_t a_{t-k-1} a_{t-k-2} + \beta^2 a_{t-1} a_{t-2} a_{t-k-1} a_{t-k-2}).$$

For all $k \neq 0$, each of the terms in the ACF has zero expectation, so that, as far as its second-order properties are concerned, η_t behaves just like an independent process. However, the MMSE forecast of a future observation, η_{t+1}, is not zero (the unconditional expectation), but is the conditional expectation

$$\hat{\eta}_{t+1} = E(\eta_{t+1} \mid \eta_t, \eta_{t-1}, \ldots) = \beta a_t a_{t-1} .$$

It then follows that x_t is not a martingale, for

$$E[x_{t+1} - x_t \mid \eta_t, \eta_{t-1}, \ldots] = \hat{\eta}_{t+1} \neq 0 ,$$

and the non-linear structure of the η_t process could be used to improve the forecasts of x_t over the simple 'no-change' forecast associated with the martingale model.

4.2 Testing the random walk hypothesis

Notwithstanding the above discussion, the random walk model has played a major role in the empirical analysis of financial time series: see, for example, the seminal research of Fama (1965) and Granger and Morgenstern (1970). In chapter 3 we examined various tests for the presence of a random walk in an observed time series. In the main these were developed by assuming that there was a specific alternative to the random walk null: for example, the stationary AR(1) process advanced by Shiller (1981a) as a model of stock market fads, and the Poterba and Summers (1988) UC model in which this AR(1) process is added to a pure random walk. There have also been numerous other tests developed against a variety of other alternatives, some of which we now discuss.

4.2.1 Autocorrelation tests

Using the results stated in chapter 2.5.1, if $w_t = \nabla x_t$ is strict white noise then the asymptotic distribution of the sample autocorrelations (standardised by $\sqrt{T}$) calculated from the realisation $\{w_t\}_1^T$ will be $N(0,1)$, so that the random walk null would be rejected at the 0.05 significance level if, for example, $\sqrt{T}|r_1| > 1.96$.

If a set of sample autocorrelations are considered, say $r_1, \ldots, r_K$, then some will probably be significant even if the null is true; on average one out of twenty would be significant at the 0.05 significance level. As noted in chapter 3.3.2, the portmanteau statistics $Q^*(K)$ and $Q(K)$ may be used in such circumstances. On the random walk null, both statistics are distributed as χ_K^2, so that the null would be rejected for sufficiently high values. Note that these tests do not require a specific alternative hypothesis: they may thus be regarded as 'diagnostic' tests with, hopefully, some power against the null for a wide range of alternatives.

The tests do, however, require that the innovations to the random walk be strict white noise. As other assumptions are possible, other types of test might be appropriate. Taylor (1986, chapter 6.4) discusses both the non-parametric runs test and certain spectral tests, which may be considered when the alternative to random walk behaviour is a cyclical model.

4.2.2 Calendar effects

As remarked above, autocorrelation tests are generally diagnostic checks aimed at detecting general departures from white noise and do not consider autocorrelations associated with specific timing patterns, i.e., patterns associated with 'calendar effects'. There has been a great deal of research carried out in recent years on detecting such effects: to date, researchers have found evidence of a January effect, in which stock returns in this month are exceptionally large when compared to the returns observed for other months; a weekend effect, in which Monday mean returns are negative rather than positive as for all other weekdays; a holiday effect, showing a much larger mean return on the day before holidays; a turn-of-the-month effect, in which the four-day return around the turn of a month is *greater* than the average total monthly return; and even an intraday effect, where prices rise sharply during the first forty-five minutes of the trading day (except Mondays!) and near the end of the day.

These 'anomalies' are reviewed in entertaining fashion by Thaler (1987a, 1987b) and some of the statistical techniques used to detect them are discussed in Taylor (1986, pp. 41–4). A wide range of techniques have been employed and discussion of them here would take us too far afield from our development of formal time series models: the interested reader is therefore recommended to examine the papers cited in the above references.

4.2.3 Consequences of non-linearities for random walk tests

As has been emphasised, the test statistics introduced above rely on the assumption that the random walk innovation is strict white noise. What happens if this innovation is just white noise, so that the sequence is merely uncorrelated rather than independent? Consider the following example, taken from Taylor (1986, p. 117). Suppose that $\{w_t\}_1^T$ is constructed from the strict white-noise process $\{a_t\}_1^T$ by

$$w_t = \begin{cases} \sigma_1 a_t, & \text{for } 1 \le t \le m, \\ \sigma_2 a_t, & \text{for } m+1 \le t \le T. \end{cases}$$

Ignoring the sample mean $\bar{w}$, the autocorrelations are given by

$$r_k = \frac{\sum_{t=1}^{T-k} w_t w_{t+k}}{\sum_{t=1}^{T} w_t^2}.$$

Keeping σ_1 fixed at a positive number and letting $\sigma_2 \to 0$, we have

$$r_k \to \frac{\sum_{t=1}^{m-k} a_t a_{t+k}}{\sum_{t=1}^{m} a_t^2},$$

which are the autocorrelations of the sequence $\{a_t\}_1^m$. These have variances of approximately m^{-1}, so that $T \cdot V(r_i) \to T/m$, which can be arbitrarily high, and certainly bigger than the value of unity that occurs when w_t is strict white noise rather than just uncorrelated, as it is here by construction.

4.3 Stochastic variance models

A simple way in which non-linearity can be introduced into a time series is to allow the variance (or the conditional variance) of the process to change, either at certain discrete points in time or continuously. Although a stationary process must have a constant variance, certain conditional variances can change: for a non-linear stationary process $\{x_t\}$, the variance, $V(x_t)$, is a constant for all t, but the conditional variance $V[x_t \mid x_{t-1}, x_{t-2}, \ldots]$ depends on the observations and thus can change from period to period.

4.3.1 Step change variance models

A simple example of a changing variance model is the following. Suppose $\{x_t\}_1^T$ (T even) is an independent sequence generated by the process

$$x_t = V_t,$$

where

$$P(V_t = \sigma_1) = P(V_t = -\sigma_1) = 0.5, \quad 1 \le t \le T/2$$
$$P(V_t = \sigma_2) = P(V_t = -\sigma_2) = 0.5, \quad T/2 < t \le T.$$

Hence

$$E(x_t) = 0, \quad \text{all } t, \quad V(x_t) = \begin{cases} \sigma_1^2, & 1 \le t \le T/2 \\ \sigma_2^2, & T/2 < t \le T \end{cases}.$$

Since $\{x_t\}_1^T$ is independent it will also be uncorrelated, and hence the sample autocorrelations $r_{i,x}$ will be small. However, the series of squares, $\{x_t^2\}_1^T$, will consist of $T/2$ consecutive observations all equal to σ_1^2 followed by $T/2$ observations all equal to σ_2^2. If $\sigma^2 = \sigma_1^2 + \sigma_2^2$, then the sample autocorrelations of x_t^2 are

$$r_{k,x^2} = \frac{\sum_{t=1}^{T-k}(x_t^2 - \sigma^2)(x_{t+k}^2 - \sigma^2)}{\sum_{t=1}^{T}(x_t^2 - \sigma^2)^2} = \frac{T-3k}{T}$$

for $1 \leq k \leq T/2$ and any variances σ_1^2 and σ_2^2. Thus, although the sample autocorrelations of x_t will be very small, those for x_t^2 can be very large.

4.3.2 Product process models

Let us now suppose that the sequence $\{x_t\}_1^T$ is generated by the *product process*

$$x_t = V_t U_t, \tag{4.2}$$

where now

$$\begin{aligned} P(U_t = 1) &= P(U_t = -1) = 0.5, \quad \text{all } t \\ P(V_t = \sigma_1) &= P(V_t = \sigma_2) \ = 0.5, \quad \text{all } t, \end{aligned} \tag{4.3}$$

and where

$$P(V_t = v_t \mid V_{t-j} = v_{t-j}, \text{all } j > 0) = \begin{cases} \alpha & \text{if } v_t = v_{t-1} \\ 1-\alpha & \text{if } v_t \neq v_{t-1} \end{cases}, \quad \alpha \neq 0.5, \tag{4.4}$$

v_t being either σ_1 or σ_2. The sequence $\{U_t\}_1^T$ is strict white noise and is independent of the sequence $\{V_t\}_1^T$. Since both are also strictly stationary, it follows that $\{x_t\}_1^T$ is also strictly stationary, with

$$E(x_t) = 0, \qquad V(x_t) = (\sigma_1^2 + \sigma_2^2)/2, \text{ all } t.$$

It is also uncorrelated as

$$E[x_t x_{t+i}] = E[V_t V_{t+i}] E[U_t] E[U_{t+i}] = 0,$$

because U_t is zero-mean strict white noise. Now

$$\begin{aligned} P(x_t = \sigma_1 \mid x_{t-1} = \sigma_1) &= P(U_t = 1) P(V_t = \sigma_1 \mid V_{t-1} = \sigma_1) \\ &= 0.5\alpha \neq P(x_t = \sigma_1) = 0.25. \end{aligned}$$

Therefore x_t and x_{t-1} are not independent random variables and, consequently, the process $\{x_t\}_1^T$ is white noise, but not strict white noise.

Consider now the conditional variance $V[x_t|x_{t-1},x_{t-2},\ldots]$. This will be given by

$$\begin{aligned} V[x_t|x_{t-1},x_{t-2},\ldots] &= E[x_t^2|x_{t-1},x_{t-2},\ldots] = E[V_t^2|x_{t-1},x_{t-2},\ldots] \\ &= \alpha x_{t-1}^2 + (1-\alpha)(\sigma_1^2+\sigma_2^2-x_{t-1}^2), \end{aligned}$$

and, since $\alpha \neq 0.5$, it will differ from the unconditional variance $V(x_t)$.

It also follows that

$$E[x_t^2] = 0.5(\sigma_1^2+\sigma_2^2), \qquad E[x_t^4] = 0.5(\sigma_1^4+\sigma_2^4),$$

and

$$\begin{aligned} E[x_t^2x_{t+1}^2] &= \alpha E[V_t^2V_{t+1}^2|V_t = V_{t-1}] + (1-\alpha)E[V_t^2V_{t+1}^2|V_t \neq V_{t-1}] \\ &= 0.5\alpha(\sigma_1^4+\sigma_2^4) + (1-\alpha)\sigma_1^2\sigma_2^2. \end{aligned}$$

Hence

$$\begin{aligned} \rho_{1,x^2} &= \frac{E[x_t^2x_{t+1}^2] - E[x_t^2]E[x_{t+1}^2]}{V[x_t^2]} = \frac{0.25(2\alpha-1)(\sigma_1^2-\sigma_2^2)^2}{0.25(\sigma_1^2-\sigma_2^2)^2} \\ &= 2\alpha - 1. \end{aligned}$$

Similarly, it can be shown that

$$\rho_{k,x^2} = (2\alpha-1)^k, \quad k>0,$$

so that, although x_t is white noise, x_t^2 has the same ACF as an AR(1) process with parameter $2\alpha-1$. In this case, however, this autocorrelation is a consequence of changes in the conditional variance of x_t, rather than any change in the unconditional variance, which remains constant.

Generalising further, suppose that x_t is still generated by the process (4.2) but now $U_t \sim NID(0,1)$ while V_t continues to follow the process given by equations (4.3) and (4.4). The process $\{x_t\}$ remains stationary and uncorrelated with zero mean and variance $0.5(\sigma_1^2+\sigma_2^2)$. It will be conditionally normal, with a conditional variance given by

$$V[x_t|v_t] = v_t^2,$$

and

$$V[x_t|v_{t-1},v_{t-2},\ldots] = \alpha v_{t-1}^2 + (1-\alpha)(\sigma_1^2+\sigma_2^2-v_{t-1}^2).$$

$E[x_t^2]$ and $E[x_t^2x_{t+i}^2]$ are as before but now $E[x_t^4] = 3(\sigma_1^4+\sigma_2^4)/2$ as $E[U_t^4]=3$. Hence

$$\begin{aligned} \rho_{i,x^2} &= \frac{(\sigma_1^2-\sigma_2^2)^2}{(\sigma_1^2-\sigma_2^2)^2 + 4(\sigma_1^4+\sigma_2^4)}(2\alpha-1)^i \\ &< (1/5)(2\alpha-1)^i, \quad i>0. \end{aligned}$$

The continuous distribution of U_t, compared with the earlier discrete distribution, induces smaller autocorrelation in x_t^2, and the ACF will have the shape of an ARMA(1,1) process. Moreover, x_t will also have high kurtosis, for

$$E[x_t^4]/E[x_t^2]^2 = 3\{1 + (\sigma_1^2 - \sigma_2^2)^2/(\sigma_1^2 + \sigma_2^2)^2\} > 3 .$$

This is a consequence of the unconditional distribution of x_t being a mixture of two normals.

The product process generating x_t given in equation (4.2) can be generalised to take the form

$$x_t = \mu + V_t U_t , \tag{4.5}$$

where $\{U_t\}$ is a standardised process, so that $E(U_t) = 0$ and $V(U_t) = 1$ for all t, and $\{V_t\}$ is a process of positive random variables usually such that $V[x_t \mid V_t] = V_t^2$: V_t is thus the conditional standard deviation of x_t.

Typically $U_t = (x_t - \mu)/V_t$ is assumed to be normal and independent of V_t: we will further assume that it is strict white noise. These assumptions together imply that the $\{x_t\}$ sequence has mean μ, variance

$$E[(x_t - \mu)^2] = E[V_t^2 U_t^2] = E[V_t^2]E[U_t^2] = E[V_t^2] ,$$

and autocovariances

$$E[(x_t - \mu)(x_{t+i} - \mu)] = E[V_t V_{t+i} U_t U_{t+i}] = E[V_t V_{t+i} U_t]E[U_{t+i}] = 0 ,$$

i.e., it is white noise. However, note that both the squared and absolute deviations, $S_t = (x_t - \mu)^2$ and $M_t = |x_t - \mu|$, can be autocorrelated. For example

$$\begin{aligned} E[(S_t - E(S_t))(S_{t+i} - E(S_t))] &= E[S_t S_{t+i}] - E[S_t]^2 \\ &= E[V_t^2 V_{t+i}^2]E[U_t^2 U_{t+i}^2] - E[V_t^2]^2 \\ &= E[V_t^2 V_{t+i}^2] - E[V_t^2]^2 , \end{aligned}$$

in which case we have

$$\rho_{k,S} = \rho_{k,V^2} \frac{V[V_t^2]}{V[S_t]} .$$

By similar arguments,

$$E[(M_t - E(M_t))(M_{t+i} - E(M_t))] = \delta^2[E[V_t V_{t+i}] - E[V_t]^2] ,$$

where $\delta = E[|U_t|] = (2/\pi)^{\frac{1}{2}}$, so that

$$\rho_{k,M} = \rho_{k,V} \frac{V[V_t]}{V[M_t]} .$$

What models are plausible for the conditional standard deviation $\{V_t\}$? Recall that it is a sequence of positive random variables, so a normal distribution is inappropriate. Since it is likely that V_t will be skewed to the right, a lognormal distribution would seem to be appropriate

$$\ln(V_t) \sim NID(\alpha, \beta^2), \quad \beta > 0.$$

It is well known that, in these circumstances

$$E[V_t^r] = \exp(r\alpha + \tfrac{1}{2}r^2\beta^2),$$

so that

$$V[V_t] = \exp(2\alpha + \beta^2)(exp(\beta^2) - 1)$$
$$V[V_t^2] = \exp(4\alpha + 4\beta^2)(exp(4\beta^2) - 1)$$
$$V[S_t] = \exp(4\alpha + 3\beta^2)(3exp(4\beta^2) - 1).$$

Hence

$$\frac{\rho_{k,S}}{\rho_{k,V^2}} = \frac{\exp(4\beta^2) - 1}{3\exp(4\beta^2) - 1} = A(\beta).$$

The ratio $A(\beta)$ increases monotonically with β from $A(0) = 0$ to the upper bound 1/3 as $\beta \to \infty$. A simple example of such a model is the AR(1) process

$$\ln(V_t) - \alpha = \phi\{\ln(V_t) - \alpha\} + \eta_t, \qquad (4.6)$$

where the innovation process $\{\eta_t\}$ is zero mean, Gaussian white noise, with variance $\beta^2(1 - \phi^2)$, and is independent of $\{U_t\}$.

4.3.3 Estimation of product process models

Consider again the product process (4.5)

$$x_t = \mu + V_t U_t,$$

with squared and absolute mean deviations

$$S_t = (x_t - \mu)^2 = V_t^2 U_t^2,$$

and

$$M_t = |x_t - \mu| = V_t |U_t|.$$

Let $\mu_V = E[V_t]$, $\sigma_V^2 = V[V_t]$ and $\delta = E[|U_t|]$, and since V_t and U_t are independent

$$E[M_t] = E[V_t]E[|U_t|] = \mu_V \delta,$$

and

$$E[S_t]=E[V_t^2]E[U_t^2]=\mu_V^2+\sigma_V^2.$$

From the realisation $\{x_t\}_1^T$, the sample averages

$$\bar{x}=T^{-1}\sum_{t=1}^{T}x_t,\quad \bar{m}=T^{-1}\sum_{t=1}^{T}|x_t-\bar{x}|,\quad \bar{s}=T^{-1}\sum_{t=1}^{T}(x_t-\bar{x})^2$$

can be calculated and hence estimates

$$\hat{\mu}_V=\bar{m}/\delta,\quad \hat{\sigma}_V^2=\bar{s}-(\bar{m}/\delta)^2$$

obtained by using $\bar{m}$ and $\bar{s}$ to estimate $E[M_t]$ and $E[S_t]$ respectively. Of course, an estimate of δ is required, but as U_t is assumed to be $NID(0,1)$, $\delta=E[|U_t|]=(2/\pi)^{\frac{1}{2}}=0.798$.

If $\ln(V_t)\sim NID(\alpha,\beta^2)$ then, since $E[V_t]=\exp(\alpha+\frac{1}{2}\beta^2)$ and $E[V_t^2]=\exp(2\alpha+2\beta^2)$

$$\beta^2=\ln\{E[V_t^2]/E[V_t]^2\}=\ln\{(\mu_V^2+\sigma_V^2)/\mu_V^2\}.$$

This suggests estimating β^2 by substituting $\hat{\mu}_V$ and $\hat{\sigma}_V^2$ for μ_V and σ_V^2, leading to

$$\hat{\beta}^2=\ln\{\delta^2\bar{s}/\bar{m}^2\}.$$

Similarly,

$$\alpha=\ln\{E[V_t]^2/E[V_t^2]^{\frac{1}{2}}\},$$

leading to

$$\hat{\alpha}=\ln\{\bar{m}^2/(\delta^2\bar{s}^{\frac{1}{2}})\}.$$

Taylor (1986, section 3.9) considers various methods for estimating the parameters of the stationary AR(1) model (4.6) assumed for $\ln(V_t)$: these require non-standard techniques and little is known about the properties of the estimators. We refer the reader to the above reference for details.

Example 4.1: Modelling the dollar/sterling exchange rate as a product process

Table 4.1 presents the SACF of the weekly changes in the dollar/sterling exchange rate, denoted x_t, originally analysed in example 2.5, the levels of which were found to be a driftless random walk in example 3.1. Also shown are the SACFs of the squared and absolute deviations, $s_t=(x_t-\bar{x})^2$ and $m_t=|x_t-\bar{x}|$. It is clear that, although x_t appears uncorrelated, s_t and m_t are certainly not, and the series is thus a candidate for modelling as a product process with $U_t\sim NID(0,1)$ and $\ln(V_t)\sim NID(\alpha,\beta^2)$.

From the sample of 470 observations, the following averages were therefore calculated

Table 4.1 *SACFs for the weekly change in the dollar/sterling exchange rate*

	Series		
r_k	x_t	s_t	m_t
1	−0.064	0.073	0.090
2	0.012	0.053	0.104
3	0.077	0.151	0.181
4	0.054	0.129	0.148
5	−0.021	0.025	0.029
6	0.003	−0.015	0.023
7	0.084	0.054	0.076
8	−0.011	0.021	0.048
9	0.013	0.066	0.107
10	0.042	0.021	0.059
11	−0.060	0.125	0.129
12	0.017	0.021	0.053
s.e.	0.046	0.046	0.046
$Q(12)$	12.35	33.71	54.72

Notes: s.e. is the asymptotic standard error of r_i. $Q(12) \sim \chi^2_{12}$ is the standard portmanteau statistic.

$$\bar{x} = -0.000917, \ \bar{s} = 0.000758, \ \bar{m} = 0.0204,$$

from which were obtained the parameter estimates

$$\hat{\mu}_V = 0.0256, \ \hat{\sigma}_V = 0.0103, \ \hat{\alpha} = -4.893, \ \hat{\beta} = 0.387, \ A(\hat{\beta}) = 0.184.$$

Hence, on taking the drift of x_t, μ, estimated as $\bar{x} = -0.000917$, to be zero, the product process for the change in the exchange rate is, with $v_t = \ln(V_t)$

$$x_t = \exp(v_t) \cdot U_t,$$

where

$$v_t \sim NID(-4.893, 0.150).$$

There is thus evidence that there is substantial non-linearity in dollar/sterling exchange rate changes. Whether this is best modelled as a product process is, however, debatable, and we return to modelling this series in later examples in this chapter, after other non-linear processes have been introduced.

4.3.4 Applications of product process models

There have been numerous applications of the various forms of the product process model, or the stochastic variance model as it is sometimes known, many of which are surveyed and discussed in Taylor (1986). Hsu (1977, 1979) has tested a single variance change model on weekly US stock returns over the period from July 1971 to August 1974, finding evidence of an increased variance from mid March 1973 onwards, this being attributed to a surge in news about the Watergate incident. Hsu (1982) and Menzefricke (1981) have given Bayesian analyses of the location of the unknown variance change-point, while Ali and Giaccotto (1982) have extended the model to incorporate up to four jumps in variance.

Various versions of product process models have been proposed, most notably by Clark (1973), Epps and Epps (1976) and Tauchen and Pitts (1983). Clark, for example, models the conditional variance of daily stock returns as a deterministic function of trading volumes, while Tauchen and Pitts develop a stochastic extension based on an economic model for the reaction of individual traders to separate items of information during the day.

Theoretically, these processes can be derived as discrete time approximations to the solution of the option valuation problem when the volatility of the underlying asset price is stochastic. Research in this vein has been carried out by, for example, Scott (1987), Wiggins (1987), Chesney and Scott (1989) and Melino and Turnbull (1990).

4.4 ARCH processes

4.4.1 Development of generalised ARCH processes

In the previous section the process determining the conditional standard deviations of x_t was assumed not to be a function of x_t. For example, for the AR(1) lognormal model of equation (4.6), V_t was dependent upon the information set $\{\eta_t, V_{t-1}, V_{t-2}, \ldots\}$. We now consider the case when the conditional standard deviations are a function of past values of x_t, i.e.

$$V_t = \mathscr{F}(x_{t-1}, x_{t-2}, \ldots).$$

A simple example is

$$\mathscr{F}(x_{t-1}) = \{\alpha_0 + \alpha_1(x_{t-1} - \mu)^2\}^{\frac{1}{2}}, \qquad (4.7)$$

where α_0 and α_1 are both positive. With $U_t \sim NID(0,1)$ and independent of V_t, x_t is then white noise and conditionally normal, i.e.

$$x_t \mid x_{t-1}, x_{t-2}, \ldots \sim NID(\mu, V_t^2),$$

so that

$$V(x_t | x_{t-1}) = \alpha_0 + \alpha_1(x_{t-1} - \mu)^2.$$

This process is stationary if, and only if, $\alpha_1 < 1$, in which case the unconditional variance is $V(x_t) = \alpha_0/(1-\alpha_1)$. The fourth moment is finite if $3\alpha_1^2 < 1$ and, if so, the kurtosis is given by $3(1-\alpha_1^2)/(1-3\alpha_1^2)$. This exceeds 3, so that the unconditional distribution of x_t is fatter tailed than the normal.

This model was first introduced by Engle (1982) and is known as a *first order autoregressive conditional heteroskedastic* [ARCH(1)] process. The popularity of the class of ARCH models can be gauged by the remark of Bollerslev, Chou and Kroner (1992), in their authoritative recent survey, that several hundred research papers using this model have appeared in the decade since its introduction!

A more convenient notation is to define $\epsilon_t = x_t - \mu = V_t U_t$ and $h_t = V_t^2$, so that the ARCH(1) model can be written as

$$\begin{aligned} &\epsilon_t | x_{t-1}, x_{t-2}, \ldots \sim NID(0, h_t) \\ &h_t = \alpha_0 + \alpha_1 \epsilon_{t-1}^2 . \end{aligned} \tag{4.8}$$

A natural extension is the ARCH(p) process, where (4.7) is replaced by

$$\mathscr{F}(x_{t-1}, x_{t-2}, \ldots, x_{t-p}) = \{\alpha_0 + \sum_{i=1}^{p} \alpha_i (x_{t-i} - \mu)^2\}^{\frac{1}{2}},$$

where $\alpha_0 > 0$ and $\alpha_i \geq 0$, $1 \leq i \leq p$. The process $\{x_t\}$ will be stationary if all the roots of the characteristic equation $\sum_{i=0}^{p} \alpha_i Z^{p-i} = 0$ lie outside the unit circle, i.e., if $\sum_{i=1}^{p} \alpha_i < 1$, in which case the unconditional variance is $V(x_t) = \alpha_0/(1 - \sum_{i=1}^{p} \alpha_i)$. In terms of ϵ_t and h_t, the conditional variance function is

$$h_t = \alpha_0 + \sum_{i=1}^{p} \alpha_i \epsilon_{t-i}^2 .$$

Detailed discussion of the ARCH(p) model, setting out further technical conditions that need not concern us here, may be found in, for example, Engle (1982), Milhøj (1985) and Weiss (1986a).

A practical difficulty with ARCH models is that, with p large, estimation (to be discussed later) will often lead to the violation of the non-negativity constraints on the α_is that are needed to ensure that the conditional variance h_t is always positive. In many early applications of the model a rather arbitrary declining lag structure was thus imposed on the α_is to ensure that these constraints were met. To obtain more flexibility, a further extension, to the *generalised* ARCH (GARCH) process, has been proposed (Bollerslev, 1986, 1988): the GARCH(q,p) process has the conditional variance function

$$h_t = \alpha_0 + \sum_{i=1}^{p} \alpha_i \epsilon_{t-i}^2 + \sum_{i=1}^{q} \beta_i h_{t-i}$$
$$= \alpha_0 + \alpha(B)\epsilon_t^2 + \beta(B)h_t, \tag{4.9}$$

where $q>0$ and $\beta_i \geq 0$, $1 \leq i \leq q$. By the usual ARMA analogue, the process $\{\epsilon_t\}$ will be stationary with unconditional variance $V(\epsilon_t) = \alpha_0(1-\alpha(1)-\beta(1))^{-1}$ if $\alpha(1)+\beta(1)<1$. By defining $\nu_t = \epsilon_t^2 - h_t$, (4.9) can be rearranged as

$$\epsilon_t^2 = \alpha_0 + \sum_{i=1}^{m} (\alpha_i + \beta_i)\epsilon_{t-i}^2 + \nu_t - \sum_{i=1}^{q} \beta_i \nu_{t-i}, \tag{4.10}$$

where $m = \max(q,p)$, thus revealing that ϵ_t^2 follows an ARMA(m,q) process with serially uncorrelated, although heteroskedastic, innovations ν_t: recall that $\epsilon_t^2 = h_t U_t^2$, so that $\nu_t = (U_t^2 - 1)h_t = (U_t^2 - 1)V_t^2$.

From (4.9) the conditional variance s periods ahead is

$$h_{t+s} = \alpha_0 + \sum_{i=1}^{n} (\alpha_i \epsilon_{t+s-i}^2 + \beta_i h_{t+s-i}) + \sum_{i=s}^{m} (\alpha_i \epsilon_{t+s-i}^2 + \beta_i h_{t+s-i}),$$

where $n = \min(m, s-1)$, so that

$$E[h_{t+s} | \epsilon_t, \epsilon_{t-1}, \ldots] = \alpha_0 + \sum_{i=1}^{n} ((\alpha_i + \beta_i) E[h_{t+s-i} | \epsilon_t, \epsilon_{t-1}, \ldots])$$
$$+ \sum_{i=s}^{m} (\alpha_i \epsilon_{t+s-i}^2 + \beta_i h_{t+s-i}). \tag{4.11}$$

On writing the unconditional variance of ϵ_t as

$$V(\epsilon_t) = E(\epsilon_t^2) = \sigma^2 = \alpha_0 \left[1 - \sum_{i=1}^{m} (\alpha_i + \beta_i)\right]^{-1},$$

(4.11) becomes

$$E[h_{t+s} | \epsilon_t, \epsilon_{t-1}, \ldots] = \sigma^2 + \sum_{i=1}^{n} ((\alpha_i + \beta_i) E[h_{t+s-i} | \epsilon_t, \epsilon_{t-1}, \ldots] - \sigma^2)$$
$$+ \sum_{i=s}^{m} (\alpha_i \epsilon_{t+s-i}^2 + \beta_i h_{t+s-i}),$$

and it follows that $E[h_{t+s} | \epsilon_t, \epsilon_{t-1}, \ldots] \rightarrow \sigma^2$ as $s \rightarrow \infty$.

As an illustration, consider the GARCH(1,1) model, where

$$h_t = \alpha_0 + \alpha_1 \epsilon_{t-1}^2 + \beta_1 h_{t-1}, \ \alpha_0 > 0, \ \alpha_1 \geq 0, \ \beta_1 \geq 0,$$

or

$$\epsilon_t^2 = \alpha_0 + (\alpha_1 + \beta_1)\epsilon_{t-1}^2 + \nu_t - \beta_1 \nu_{t-1}.$$

Stationarity is ensured if $\alpha_1+\beta_1<1$ and, for $s\geq 2$

$$E[h_{t+s}|\epsilon_t,\epsilon_{t-1},\ldots]=\alpha_0+(\alpha_1+\beta_1)E[h_{t+s-1}|\epsilon_t,\epsilon_{t-1},\ldots].$$

Suppose now that $\alpha_1+\beta_1=1$ and $\alpha_0=0$: this leads to the *integrated* GARCH, or IGARCH(1,1) model (Engle and Bollerslev, 1986)

$$h_t=\alpha_1\epsilon_{t-1}^2+(1-\alpha_1)h_{t-1}.$$

Thus

$$\begin{aligned}E[h_{t+s}|\epsilon_t,\epsilon_{t-1},\ldots]&=E[h_{t+s-1}|\epsilon_t,\epsilon_{t-1},\ldots]=\ldots\\&=E[h_{t+1}|\epsilon_t,\epsilon_{t-1},\ldots]=h_{t+1}.\end{aligned}$$

The conditional variance s steps ahead is the same as the conditional variance one-step ahead for all horizons s, and the model is obviously closely related to the traditional random walk, which has a unit root in the conditional mean rather than, as here, a unit root in the conditional variance. When $\alpha_0>0$, there is also a trend in the conditional variance, so that

$$E[h_{t+s}|\epsilon_t,\epsilon_{t-1},\ldots]=s\alpha_0+h_{t+1}.$$

In both cases, the effect of h_{t+1} on the future conditional variances is permanent: shocks are not forgotten.

In the basic ARCH model (4.8), the distribution of ϵ_t was assumed to be conditionally normal. Bollerslev (1987) considers the case when the distribution is standardised t with unknown degrees of freedom υ that may be estimated from the data: for $\upsilon>4$ such a distribution is leptokurtic and hence has 'fatter tails' than the normal distribution. Other distributions that have been considered include the normal–Poisson mixture distribution in Jorion (1988), the power exponential distribution in Baillie and Bollerslev (1989), the normal–lognormal mixture in Hsieh (1989a), and the generalised exponential distribution in Nelson (1991).

A second modification is to allow the relationship between h_t and ϵ_t to be more flexible than the quadratic mapping that has so far been assumed. Engle and Bollerslev (1986) consider two variants to the GARCH(1,1) model

$$h_t=\alpha_0+\beta_1 h_{t-1}+\alpha_1|\epsilon_{t-1}|^{\lambda_1}, \tag{4.12}$$

and

$$h_t=\alpha_0+\beta_1 h_{t-1}+\alpha_1(2F(\epsilon_{t-1}^2/\lambda_2)-1), \tag{4.13}$$

where F denotes the standard normal cumulative distribution function. If $\lambda_1=2$ in (4.12) then the model is just the GARCH(1,1), while if, in (4.13), λ_2 is large relative to σ, the model is essentially linear. Geweke (1986) and

Pantula (1986) suggest a logarithmic parameterisation, while Higgins and Bera (1992) generalise further by considering a specification in which both h_t and the ϵ_ts are transformed using the Box and Cox (1964) power transformation.

Nelson (1991) and Pagan and Schwert (1990) consider another variant, the *exponential* GARCH (EGARCH) model, in which h_t is an asymmetric function of past ϵ_ts. The EGARCH(1,1) model is

$$\ln(h_t)=\alpha_0+\beta_1\ln(h_{t-1})+\alpha_1(\theta U_{t-1}+(|U_{t-1}|-(2/\pi)^{\frac{1}{2}})),$$

where $U_t=\epsilon_t/h_t$, and the general EGARCH(p,q) model is discussed in more detail in chapter 5.

An alternative approach is, instead of relying on a parametric representation for h_t, to use a non-parametric estimation technique to approximate the conditional variance: see Bollerslev, Chou and Kroner (1992) for a discussion and references.

4.4.2 Estimation of ARMA models with ARCH errors

The analysis so far has proceeded on the assumption that $\epsilon_t=x_t-\mu$ is serially uncorrelated. A natural extension is to allow x_t to follow an ARMA(P,Q) process, so that the combined ARMA–ARCH model becomes

$$\phi(B)(x_t-\mu)=\theta(B)\epsilon_t, \tag{4.14}$$

$$h_t=E[\epsilon_t^2\,|\,\epsilon_{t-1},\epsilon_{t-2},\ldots]=\alpha_0+\sum_{i=1}^{p}\alpha_i\epsilon_{t-i}^2+\sum_{i=1}^{q}\beta_i h_{t-i}. \tag{4.15}$$

This latter equation can be written as

$$h_t=z_t'\omega=z_{1t}'\omega_1+z_{2t}'\omega_2,$$

where

$$z_t'=(z_{1t}':z_{2t}')=(1,\epsilon_{t-1}^2,\ldots,\epsilon_{t-p}^2:h_{t-1},\ldots,h_{t-q}),$$

and

$$\omega'=(\omega_1':\omega_2')=(\alpha_0,\alpha_1,\ldots,\alpha_p:\beta_1,\ldots,\beta_q).$$

Using this notation, maximum likelihood (ML) estimates of the model can be obtained in the following way. Define Θ as the vector of parameters in the model given by equations (4.14) and (4.15) and partition it as $\Theta=(\omega':\psi')$; $\psi'=(\phi_1,\ldots,\phi_P,\theta_1,\ldots,\theta_Q,\mu)$ being a vector containing the parameters in the ARMA equation. We may also define $\Theta_0=(\omega_0':\psi_0')$ as the true parameter vector.

The log likelihood function for a sample of T observations is, apart from some constants,

$$L_T(\Theta)=T^{-1}\sum_{t=1}^{T} l_t(\Theta),$$

where

$$l_t(\Theta)=-\tfrac{1}{2}(\log(h_t)+\epsilon_t^2 h_t^{-1}).$$

Precise details of ML estimation may be found in, for example, Engle (1982), Weiss (1986a, 1986b) and Bollerslev (1988). The Berndt, Hall, Hall and Hausman (BHHH, 1974) algorithm is a convenient method of computation. If $\Theta^{(i)}$ denotes the parameter estimates after the ith iteration, then $\Theta^{(i+1)}$ is calculated by the algorithm as

$$\Theta^{(i+1)}=\Theta^{(i)}+\lambda_i\left(\sum_{t=1}^{T}\frac{\partial l_t}{\partial\Theta}\frac{\partial l_t}{\partial\Theta'}\right)^{-1}\sum_{t=1}^{T}\frac{\partial l_t}{\partial\Theta}, \qquad (4.16)$$

where $\partial l_t/\partial\Theta$ is evaluated at $\theta^{(i)}$ and λ_i is a variable step length chosen to maximise the likelihood function in the given direction. Because the information matrix, $\mathfrak{Z}=-E(\partial^2 l_t/\partial\Theta\partial\Theta')$, is block diagonal, ω can be estimated without loss of asymptotic efficiency based on a consistent estimate of ψ, and vice versa, so that the iterations for $\omega^{(i)}$ and $\psi^{(i)}$ can be carried out separately.

The ML estimate $\hat{\Theta}_{ML}$ is strongly consistent for Θ_0 and asymptotically normal with mean Θ_0 and covariance matrix $\mathfrak{Z}^{-1}$, consistently estimated by $T^{-1}(\sum_{t=1}^{T}(\partial l_t/\partial\Theta)(\partial l_t/\partial\Theta'))^{-1}$, which may be obtained from the last BHHH iteration.

As $E(\epsilon_t^2)=V(\epsilon_t)=\sigma^2$ is constant and finite, LS estimates of the ARMA parameters, $\hat{\psi}_{LS}$, are found by minimising $\hat{\sigma}^2(\psi)=T^{-1}\sum\epsilon_t^2$ in the usual way. These will be consistent and asymptotically normal if $E(\epsilon_t^4)$ exists. Although the actual form of ARCH is not needed for obtaining $\hat{\psi}_{LS}$, the covariance matrix associated with these estimates is given by $C=A^{-1}BA^{-1}$, where

$$A=2E[(\partial\epsilon_t/\partial\psi)(\partial\epsilon_t/\partial\psi')]$$

and

$$B=4E[\epsilon_t^2(\partial\epsilon_t/\partial\psi)(\partial\epsilon_t/\partial\psi')],$$

both evaluated at ψ_0. LS estimates of the ARCH parameters are obtained by estimating the regression (4.10) with the squared residuals from the ARMA model, e_t^2, being used in place of ϵ_t^2. Again, adjustments are required to the associated covariance matrix: the appropriate expression is found in Weiss (1986b).

4.4.3 *Testing for the presence of ARCH errors*

Let us suppose that an ARMA model for x_t has been estimated, from which the residuals e_t have been obtained. The presence of ARCH can lead to serious model misspecification if it is ignored: as with all forms of heteroskedasticity, analysis assuming its absence will result in inappropriate parameter standard errors, and these will typically be too small. For example, Weiss (1984) shows that ignoring ARCH will lead to the identification of ARMA models that are overparameterised.

Methods for testing whether ARCH is present are therefore essential, particularly as estimation incorporating it requires expensive iterative techniques. Equation (4.10) has shown that if ϵ_t is GARCH(q,p), then ϵ_t^2 is ARMA(m,q), where $m=\max(q,p)$, and Bollerslev (1986) shows that standard ARMA theory follows through in this case. This implies that the squared residuals e_t^2 can then be used to identify m and q, and therefore p, in a fashion similar to the way the usual residuals are used in conventional ARMA modelling. McLeod and Li (1983), for example, show that the sample autocorrelations of e_t^2 have asymptotic variance T^{-1} and that portmanteau statistics calculated from them are asymptotically χ^2 if the ϵ_t^2 are independent.

Formal tests are also available. Engle (1982) shows that a test of the null hypothesis that ϵ_t has a constant conditional variance against the alternative that the conditional variance is given by an ARCH(p) process, i.e., a test of $\alpha_1=\ldots=\alpha_p=0$ in (4.15), conditional upon $\beta_1=\ldots=\beta_q=0$, may be based on the Lagrange Multiplier principle. The test procedure is to run a regression of e_t^2 on a constant and $e_{t-1}^2, e_{t-2}^2, \ldots, e_{t-p}^2$ and to test the statistic $T\cdot R^2$ as a χ_p^2 variate, where R^2 is the squared multiple correlation coefficient of the regression.

When the alternative is a GARCH(q,p) process, some complications arise. In fact, a general test of $q>0$, $p>0$ against a white-noise null is not feasible, nor is a test of GARCH($q+r_1$,$p+r_2$) errors, where $r_1>0$ and $r_2>0$, when the null is GARCH(q,p). Furthermore, under this null, the LM test for GARCH(r,p) and ARCH($p+r$) alternatives coincide. What can be tested is the null of an ARCH(p) process against a GARCH(q,p) alternative, i.e., to test H_0:$\omega_2=0$. Bollerslev (1988) shows that the appropriate test statistic is $T\cdot R^2$ computed from the regression of $(\epsilon_t^2 h_t^{-1}-1)$ on the vector $(h_t \partial h_t/\partial\omega)$, both being evaluated under H_0: i.e., the regressors are $(h_t z_t)$.

4.4.4 *A further extension of ARCH models*

Weiss (1986b) considers a further extension of the ARCH process. Rather than restricting h_t to be just a function of ϵ_t alone, past values of x_t are allowed to appear as well, so that, along with the ARMA model (4.14) for x_t, we now have the conditional variance equation

$$h_t = \alpha_0 + \sum_{i=1}^{p} \alpha_i \epsilon_{t-i}^2 + \delta_0(\hat{x}_t - \mu)^2 + \sum_{i=1}^{r} \delta_i (x_{t-i} - \mu)^2, \tag{4.17}$$

where $\hat{x}_t = E(x_t | x_{t-1}, x_{t-2}, \ldots)$ is the forecast of x_t made at $t-1$. Estimation and testing of this model follows the procedures outlined above with appropriate modifications.

4.4.5 ARCH and theories of asset pricing

The importance of ARCH processes in modelling financial time series is seen most clearly in models of asset pricing which involve agents maximising expected utility over uncertain future events. To illustrate this, consider the following example, taken from Engle and Bollerslev (1986). Suppose a representative agent must allocate his wealth, W_t, between shares of a risky asset q_t at a price p_t and those of a risk-free asset x_t, whose price is set equal to 1. The shares of the risky asset will be worth y_{t+1} each at the end of the period (if there are no dividends, then $y_{t+1} = p_{t+1}$). The risk-free asset will be worth $r_t x_t$, where r_t denotes one plus the risk-free rate of interest. If the agent has a mean–variance utility function in end-of-period wealth, $W_{t+1} = q_t y_{t+1} + r_t x_t$, then the allocation problem for the agent is to maximise this utility function with respect to holdings of the risky asset, q_t, i.e., to maximise

$$2E_t(q_t y_{t+1} + r_t x_t) - \gamma_t V_t(q_t y_{t+1}),$$

subject to the start of period wealth constraint

$$W_t = x_t + p_t q_t.$$

This has the solution

$$p_t = r_t^{-1} E_t(y_{t+1}) - \gamma_t q_t r_t^{-1} V_t(y_{t+1}). \tag{4.18}$$

If the outstanding stock of the risky asset is fixed at q, and γ_t and r_t are taken as constants (γ and r, respectively), then (4.18) describes the asset pricing model.

If the risky asset is interpreted as a forward contract for delivery in s periods time, the price that a pure speculator would be willing to pay is

$$p_t = r^{-s}[E_t(y_{t+s}) - \delta V_t(y_{t+s})], \tag{4.19}$$

where r^{-s} gives the present discounted value at the risk-free rate r and $\delta = \gamma q$. A simple redating of the model shows that the price of the forward contract at time $t+1$, for $s \geq 2$ periods remaining to maturity, can be expressed as

$$p_{t+1} = r^{1-s}[E_{t+1}(y_{t+s}) - \delta V_{t+1}(y_{t+s})].$$

Taking expectations at time t, multiplying by r^{-1}, and subtracting from (4.19) gives

$$p_t = r^{-1}E_t(p_{t+1}) - \delta r^{-s}[V_t(y_{t+s}) - E_t(V_{t+1}(y_{t+s}))]. \tag{4.20}$$

Now, suppose y_t can be represented by an infinite moving average process where the innovations are uncorrelated but have time-varying conditional variance h_t

$$\begin{aligned} y_t &= \epsilon_t + \sum_{i=1}^{\infty} \theta_i \epsilon_{t-i} = \theta(B)\epsilon_t, \\ V_t(y_{t+1}) &= V_t(\epsilon_{t+1}) = h_{t+1}. \end{aligned} \tag{4.21}$$

Thus

$$V_t(y_{t+s}) = E_t\left(\sum_{i=1}^{s} \theta_{s-i}\epsilon_{t+i}\right)^2 = \sum_{i=1}^{s} \theta_{s-i}^2 E_t(h_{t+i}).$$

Consequently,

$$V_t(y_{t+s}) - E_t(V_{t+1}(y_{t+s})) = \theta_{s-1}^2 h_{t+1},$$

and (4.20) becomes

$$p_t = r^{-1}E_t(p_{t+1}) - \delta r^{-s}\theta_{s-1}^2 h_{t+1},$$

which is the familiar formula for a one-period holding yield with the explicit calculation of the effect of the changing variance of y_{t+s} for a risk averse agent.

In this simple model the only source of uncertainty derives from the future spot price to which the contract relates. In many other situations, however, there is a flow of uncertain distributions which accrue to the owner of the asset: for example, the price of a share is determined by the present discounted value of the expected dividend stream. The precise form in which the variability of future payoffs enters the asset pricing formulation will depend, amongst other things, on the utility function of the agents and the intertemporal substitutability of the payouts. A simple formulation might be

$$p_t = \sum_{s=1}^{\infty} r^{-s}[E_t(y_{t+s}) - \delta V_t(y_{t+s})],$$

where $\{y_t\}_{t+1}^{\infty}$ is the future income stream generated by the asset. If y_t again follows the process (4.21), this pricing equation can be converted to the holding yield expression

$$p_t = r^{-1}[E_t(p_{t+1}) + E_t(y_{t+1}) - \delta\lambda h_{t+1}],$$

where λ depends upon $\theta(B)$ and r.

It is thus clear that, if $\delta \neq 0$, the conditional variance of y_t in the future will affect the price of the asset today. If such variances can be forecast as in equation (4.11), then the current information on y_t and the current conditional variance will have an effect on the current price. The size of the effect, however, will depend upon the persistence of the variance, i.e., on how important current information is in predicting future variances.

A closed form solution to the simple asset pricing formula (4.19) depends upon the process assumed to generate the 'forcing variable' y_t. Suppose y_t is a random walk with innovations that follow an integrated GARCH(1,1) process. Thus $E_t(y_{t+s}) = y_t$ and

$$V_t(y_{t+s}) = E_t\left(\sum_{i=1}^{s} \epsilon_{t+i}^2\right) = E_t\left(\sum_{i=1}^{s} h_{t+i}\right) = sh_{t+1},$$

so that

$$p_t = r^{-s}[y_t - \delta s h_{t+1}].$$

For a future contract where no money changes hands until the terminal date $t+s$, the risk-free rate of return is zero so that $r=1$, i.e., the solution simplifies to

$$p_t = y_t - \delta s h_{t+1}.$$

If $\delta \neq 0$ there will be a time-varying risk premium in the future contract. For contracts far in the future, new information will have a substantial effect on asset prices as it changes agents' perceptions of the variance of the final payoff as well as all the intermediate variances. This persistence gives time-varying risk premia even for contracts many periods into the future and thus implies sizeable effects on asset prices.

Alternatively, suppose that the random walk innovations to y_t are serially independent with constant variance σ^2. In this case $V_t(y_{t+s}) = s\sigma^2$ and the solution to (4.19) is

$$p_t = y_t - \delta s \sigma^2,$$

so that, although the variance of the spot price enters the pricing equation, it does not give rise to a time-varying risk premium since new information casts no light on future uncertainty.

Finally, consider an intermediate case when the innovations are GARCH(1,1) such that $\alpha_1+\beta_1<1$. The unconditional variance will be $\sigma^2=\alpha_0/(1-\alpha_1-\beta_1)$ and it follows from (4.11) that

$$E_t(h_{t+s}-\sigma^2)=(\alpha_1+\beta_1)^{s-1}(h_{t+1}-\sigma^2),$$

and

$$V_t(y_{t+s})=\sum_{i=1}^{s}(\sigma^2+E_t(h_{t+i}-\sigma^2))=s\sigma^2+(h_{t+1}-\sigma^2)\left(\frac{1-(\alpha_1+\beta_1)^s}{1-\alpha_1-\beta_1}\right).$$

Substituting into (4.19), the solution of the future contract is

$$p_t=y_t-\delta s\sigma^2+\delta(h_{t+1}-\sigma^2)\left(\frac{1-(\alpha_1+\beta_1)^s}{1-\alpha_1-\beta_1}\right).$$

Current information, embodied in the term $h_{t+1}-\sigma^2$, continues to be an important part of the time-varying risk premium even for large s but, in contrast to the solution for the integrated GARCH(1,1) model, where $\alpha_1+\beta_1=1$, its relative importance decreases with the length of the contract.

These examples thus establish that a solution to an asset pricing equation depends in a crucial way on the distribution of the forcing variable, y_t, in particular on its conditional variance, which is naturally modelled as an ARCH process.

We should also note that, analogous to product processes being discrete approximations to continuous time option valuation models, ARCH models can approximate a wide range of stochastic differential equations: see Nelson (1990). Indeed, Nelson also shows that the EGARCH(0,1) process and the product process of (4.5) and (4.6) are discrete time approximations to the same continuous time model. Further analysis of the predictive aspects of ARMA–ARCH models is developed in Baillie and Bollerslev (1992).

While serial correlation in conditional variances is clearly a property of asset prices, a systematic search for its causes has only recently begun. A popular explanation for the prominence of ARCH effects is the presence of a serially correlated news arrival process (see, for example, Diebold and Nerlove, 1989) and Engle, Ito and Lin (1990) find a measure of empirical support for this hypothesis.

Bollerslev, Chou and Kroner (1992) survey the literature on the foundation of ARCH models and, in particular, focus on the application of ARCH models to stock return and interest rate data, emphasising the use of ARCH to model volatility persistence, and to foreign exchange rate data, where the characterisation of exchange rate movements have important implications for many issues in international finance.

Example 4.2: Modelling the dollar/sterling exchange rate as an ARCH process

Since the changes in the dollar/sterling exchange rate are uncorrelated, but the squared changes are correlated (see table 4.1), an alternative possibility for modelling this series is an ARCH process. Formal tests for the presence of ARCH are provided by the Q statistics in table 4.1 (noting that the sample mean $\bar{x}$ is very small and statistically insignificant, so that x_t^2 and s_t are almost identical), and the LM test statistic $T \cdot R^2$ from the regression of x_t^2 on lagged values of itself. Using $p=4$ lags, for example, yields the statistic $T \cdot R^2 = 18.71 \sim \chi_4^2$, clearly indicating the presence of ARCH.

Since the SACF dies out fairly slowly, a GARCH(1,1) process seems a reasonable candidate for modelling x_t^2. Estimation of such a process obtains

$$\underset{}{x_t^2} = \underset{(9.04\cdot 10^{-4})}{7.84\cdot 10^{-4}} + \underset{(0.327)}{0.921} x_{t-1}^2 + \nu_t - \underset{(0.332)}{0.871} \nu_{t-1}, \quad L=2359.30,$$

asymptotic standard errors being shown in parentheses.

From the relationship between (4.9) and (4.10), this can equivalently be written as

$$h_t = 7.84\cdot 10^{-4} + 0.921 x_{t-1}^2 - 0.871(x_{t-1}^2 - h_{t-1}).$$

Since $\alpha_1 + \beta_1 = 0.921$ is close to one, multi-step forecasts from the model will approach the unconditional variance of $\sigma^2 = 0.010$ quite slowly: the estimated mean lag of this variance expression, $1/(1-\beta_1)$, equals 7.75, or about eight weeks.

Reestimating the GARCH(1,1) model by imposing the restriction for integration in variance, $\alpha_1 + \beta_1 = 1$, but no trend in variance, $\alpha_0 = 0$, yields

$$\nabla x_t^2 = \nu_t - \underset{(0.046)}{0.946} \nu_{t-1}, \quad L=2357.17,$$

or

$$h_t = 0.054\epsilon_{t-1}^2 + 0.946 h_{t-1}.$$

A comparison of the two log likelihoods implies a likelihood ratio test statistic of 4.26 with two degrees of freedom. There are, of course, complications involved in such tests, as the null hypothesis lies on the boundary of the parameter space, but it certainly seems plausible to conclude that the change in the dollar/sterling exchange rate, although serially uncorrelated, has a time-varying conditional variance which can be modelled as an integrated GARCH process.

This conclusion is, in fact, consistent with the many studies of exchange rates that find ARCH models to provide a satisfactory representation of the dynamic behaviour of such series: see, for example, Milhøj (1987), Diebold and Nerlove (1989), Hsieh (1988, 1989a, 1989b) and Baillie and Bollerslev (1989).

4.5 Other non-linear univariate models

4.5.1 Bilinear processes

An important class of non-linear model is the *bilinear*, which takes the general form

$$\phi(B)(x_t - \mu) = \theta(B)\epsilon_t + \sum_{i=1}^{R}\sum_{j=1}^{S}\gamma_{ij}x_{t-i}\epsilon_{t-j}, \qquad (4.22)$$

where $\epsilon_t \sim SWN(0,\sigma^2)$. The second term on the right-hand side of (4.22) is a bilinear form in ϵ_{t-j} and x_{t-i}, and this accounts for the non-linear character of the model: if all the γ_{ij} are zero, (4.22) reduces to the familiar ARMA model (4.14).

Little analysis has been carried out on this general bilinear form, but Granger and Andersen (1978) have analysed the properties of several simple bilinear forms, characterised as

$$x_t = \epsilon_t + \gamma_{ij}x_{t-i}\epsilon_{t-j}.$$

If $i > j$ the model is called superdiagonal, if $i = j$ it is diagonal, and if $i < j$, it is subdiagonal. If we define $\lambda = \gamma_{ij}\sigma$, then for superdiagonal models, x_t has zero mean and variance $\sigma^2/(1-\lambda^2)$, so that $|\lambda| < 1$ is a necessary condition for stability. Conventional identification techniques using the SACF of x_t would identify this series as white noise, but Granger and Andersen show that, in theory at least, the SACF of the squares of x_t would identify x_t^2 as an ARMA(i,j) process, so that we could distinguish between white noise and this bilinear process by analysing x_t^2.

Diagonal models will also be stationary if $|\lambda| < 1$. If $i = j = 1$, x_t will be identified as MA(1), with $0 < \rho_1 \leq 0.1547$ (corresponding to $\lambda = \pm 0.605$), while x_t^2 will be identified as ARMA(1,1). However, if x_t actually was MA(1), then x_t^2 will also be MA(1), so that this result allows the bilinear model to be distinguished from the linear model. In general, the levels of a diagonal model will be identified as MA(i).

Subdiagonal models are essentially similar to superdiagonal models in that they appear to be white noise but generally have x_t^2 following an ARMA(i,j) process. Detailed analysis of the properties of bilinear models can be found in Granger and Andersen (1978), Subba Rao (1981) and Subba Rao and Gabr (1984).

4.5.2 A comparison of ARCH and bilinearity

Weiss (1986b) provides a detailed comparison of the ARMA–ARCH model, given by equations (4.14) and (4.17), with the *bilinear* model (4.22). At first sight, the models appear quite different: whereas the addition of the ARCH equation to the pure ARMA process (4.14) introduces non-linearity by affecting the conditional variance, the addition of the bilinear terms contained in (4.22) changes the *conditional mean* of x_t. Weiss argues that, despite these different influences, the two processes can have similar properties and, for example, the bilinear process may be mistaken for an ARMA model with ARCH errors.

Why might this be? Suppose the true model for x_t is (4.22) but the ARMA model

$$\tilde{\phi}(B)(x_t - \tilde{\mu}) = \tilde{\theta}(B)\tilde{\epsilon}_t \tag{4.23}$$

is fitted. The residual $\tilde{\epsilon}_t$ is given by

$$\tilde{\epsilon}_t = \vartheta_1(B)\epsilon_t + \vartheta_2(B) \sum_{i=1}^{R} \sum_{j=1}^{S} \gamma_{ij} x_{t-i}\epsilon_{t-j},$$

where $\vartheta_1(B) = \phi^{-1}(B)\tilde{\theta}^{-1}(B)\tilde{\phi}(B)\theta(B)$ and $\vartheta_2(B) = \tilde{\phi}^{-1}(B)\tilde{\theta}^{-1}(B)\phi(B)$. On squaring this expression and taking conditional expectations, it is clear that $E[\tilde{\epsilon}_t^2 \mid x_{t-1}, x_{t-2}, \ldots]$ is not constant but will be a function of lagged ϵ_t^2, and hence may be thought to have ARCH. For example, suppose the true model is

$$x_t = \epsilon_t + \gamma_{21} x_{t-1}\epsilon_{t-1}. \tag{4.24}$$

As $E(x_t) = 0$ and $E(x_t x_{t+i}) = 0$, $i > 0$, the use of traditional modelling techniques may identify the trivial ARMA model $x_t = \tilde{\epsilon}_t$, where

$$\tilde{\epsilon}_t = \epsilon_t + \gamma_{21}\epsilon_{t-1}\tilde{\epsilon}_{t-1}.$$

Squaring this and taking expectations gives

$$E[\tilde{\epsilon}_t^2 \mid x_{t-1}, x_{t-2}, \ldots] = \sigma_\epsilon^2 + \gamma_{21}^2 \sigma_\epsilon^2 \tilde{\epsilon}_{t-1}^2.$$

Now, the LM statistic for testing whether $\tilde{\epsilon}_t$ is ARCH(1) is $T \cdot R^2$ from the regression of $\tilde{\epsilon}_t^2$ on a constant and $\tilde{\epsilon}_{t-1}^2$: given the above expectation, such a statistic may well be large even if the correct model is really the bilinear process (4.24).

The correct LM statistic for testing $x_t = \tilde{\epsilon}_t$ against the bilinear alternative

(4.24) is, in fact, $T \cdot R^2$ from the regression of $\tilde{\epsilon}_t$ on a constant, $\tilde{\epsilon}_{t-1}$ and $\tilde{\epsilon}^2_{t-1}$. In general, if $\phi(B)$ and $\theta(B)$ in (4.22) are of orders P and Q respectively, then the LM statistic for testing (4.22) against the simple linear ARMA specification (4.14) is $T \cdot R^2$ from the regression of $\tilde{\epsilon}_t$ on a constant, $x_{t-1}, \ldots, x_{t-P}, \tilde{\epsilon}_{t-1}, \ldots, \tilde{\epsilon}_{t-Q}$, and $x_{t-i}\tilde{\epsilon}_{t-j}$, $i = 1, \ldots, R, j = 1, \ldots, S$; the statistic being distributed as χ^2_{RS}. Weiss shows, however, that such a test will not have the correct size if, in fact, ARCH is present as well: nor, indeed, will the LM test for ARCH if bilinearity is present.

Weiss (1986b) shows that LS and ML estimates of the bilinear model (4.22) coincide. However, although estimation of a bilinear model is straightforward, identification of that model can pose difficulties, particularly when, as we have seen, both bilinearity and ARCH are present and one can be confused with the other.

Weiss thus considers the combined bilinear model with ARCH errors, i.e., the bilinear process (4.22) with the ARCH error specification (4.17). The identification of this model is based on the relative difficulties of the different specification errors. First, ignoring bilinearity can lead to residuals that appear to have ARCH even though they may not be autocorrelated. On the other hand, misspecifying the ARCH will affect the variance of a process but not the specification of the mean equation. Given the greater complexity of bilinear models and the difficulties in their specification, this suggests that it is easier to mistake bilinearity for ARCH than ARCH for bilinearity. Weiss thus suggests that the bilinear model should be specified before ARCH is considered explicitly.

The suggested procedure is to use the SACFs of x_t^2, $\tilde{\epsilon}_t$ and $\tilde{\epsilon}_t^2$ and associated LM tests to specify the bilinear process after a pure ARMA model has been identified and fitted by conventional techniques: the SACFs, which do not allow for ARCH, will suggest possible bilinear specifications or extra bilinear terms, and the formal tests, which do allow for ARCH, can then be used to determine which specifications are appropriate. However, because we wish to test for bilinearity in the possible presence of ARCH, the LM test, although not requiring the actual form of ARCH, no longer has a $T \cdot R^2$ representation: the exact form is given in Weiss (1986b).

Once the bilinearity has been determined, the ARCH equation can be specified using the ACF of the squared residuals obtained from the estimation of the bilinear model. Estimation of the combined model then follows, and overfitting and LM tests for extra ARCH or bilinear parameters can be attempted.

Since the LM test for bilinearity in the presence of ARCH does not have the usual $T \cdot R^2$ form, and because the subsequent ARCH test requires first estimating a bilinear model, this procedure is rather burdensome if we simply want a simple test for non-linearity which is sensitive to both ARCH and bilinear alternatives. Higgins and Bera (1988) thus propose an easily computed simultaneous test for a joint ARCH and bilinear alternative. This is an LM test whose construction exploits the result that the individual LM tests for ARCH and bilinearity are additive: the joint test statistic is thus the sum of the individual test statistics. Moreover, because the two forms of non-linearity are considered simultaneously, the LM test for bilinearity again has a standard $T \cdot R^2$ representation, being the test outlined above. Hence the combined test statistic will be distributed as χ^2_{RS+P}.

Maravall (1983) considers an alternative form of bilinearity in which x_t is given by the ARMA process

$$\phi(B)(x_t - \mu) = \theta(B)a_t,$$

but where the *uncorrelated* sequence $\{a_t\}$ is bilinear in a_t and the strict white-noise sequence $\{\epsilon_t\}$

$$a_t = \epsilon_t + \sum_{i=1}^{R} \sum_{j=1}^{S} \gamma_{ij} a_{t-i} \epsilon_{t-j}. \tag{4.25}$$

This may be interpreted as a bilinear model 'forecasting white noise'.

How useful are bilinear models in modelling financial time series? De Gooijer (1989) presents evidence to suggest that such processes can provide useful models for certain daily stock return series, although the residual variance of the bilinear models were usually only marginally smaller than those obtained from alternative linear models.

Example 4.3: Is the dollar/sterling exchange rate bilinear?

Given the above discussion, is it possible that the ARCH model fitted to the dollar/sterling exchange rate in example 4.2 is a misspecification and the true process generating the series is of bilinear form? An obvious way to proceed is to consider the SACFs of the changes and squared changes of the series shown in table 4.1 in the light of the above results on bilinear processes. Since the changes, x_t, are indistinguishable from white noise, this would appear to rule out all diagonal models, which would identify x_t as an MA process. The SACF of the squared changes have a rather complicated pattern, the significant autocorrelations appearing at lags 3 and 4. This would not seem to identify any simple sub- or superdiagonal bilinear model and makes the construction of LM tests, which require a bilinear specification to be proposed, somewhat difficult. We therefore remain unconvinced that the series is generated by a bilinear process, preferring the ARCH specification arrived at in example 4.2.

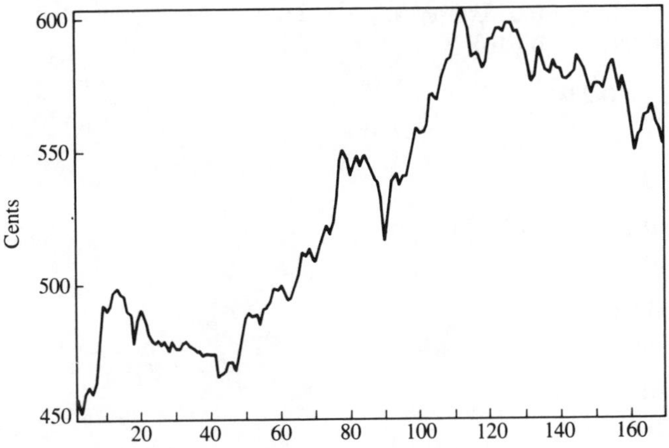

Figure 4.1 IBM common stock price (daily from 17 May 1961)

Example 4.4: Modelling IBM stock prices as a combined bilinear and ARCH process

The daily closing price for IBM common stock for the 169 trading days starting 17 May 1961, presented as part of series B in Box and Jenkins (1976) and plotted as figure 4.1, has been investigated by various researchers. Conventional (linear) identification procedures suggest that the differences of the series, denoted x_t, follow an MA(1) process, LS estimation of which obtains

$$x_t = \tilde{\epsilon}_t - 0.26\tilde{\epsilon}_{t-1}, \quad \hat{\sigma}^2 = 24.8, \quad r_{1,\tilde{\epsilon}} = -0.02, \quad r_{1,\tilde{\epsilon}^2} = 0.18.$$

An LM test for ARCH(1) errors, ignoring the possibility of bilinearity, yields a $T \cdot R^2$ statistic of 15.1, which is clearly significant (note that this confirms the evidence of non-linearity provided by the lag one autocorrelation of $\tilde{\epsilon}^2$). Incorporating such an error specification yields, on ML estimation,

$$\underset{}{x_t = \epsilon_t - \underset{(0.08)}{0.24}\epsilon_{t-1}}, \quad \hat{\sigma}^2 = 24.8, \quad r_{1,\epsilon} = 0.01, \quad r_{1,\epsilon^2} = 0.02$$

$$h_t = \underset{(2.6)}{17.87} + \underset{(0.12)}{0.28}\epsilon^2_{t-1},$$

showing that a small error implies that the variance of the next error will be small and hence the next error is more likely to be small. LM tests for first-order (diagonal) bilinearity of either the form (4.22) or (4.25) are significant, however, leading to the estimated models

$$x_t = \epsilon_t - \underset{(0.08)}{0.24}\epsilon_{t-1} + \underset{(0.010)}{0.023}x_{t-1}\epsilon_{t-1}, \quad \hat{\sigma}^2 = 23.7$$

$$h_t = \underset{(2.6)}{18.31} + \underset{(0.11)}{0.23}\epsilon_{t-1}^2,$$

and

$$x_t = a_t - \underset{(0.08)}{0.23}a_{t-1}, \quad \hat{\sigma}^2 = 23.7$$

$$a_t = \epsilon_t + \underset{(0.010)}{0.023}a_{t-1}\epsilon_{t-1}$$

$$h_t = \underset{(2.6)}{18.40} + \underset{(0.11)}{0.23}a_{t-1}^2.$$

The bilinear terms are significant and their introduction decreases the estimate of the ARCH parameter somewhat. An identical specification is arrived at if a bilinear process is first fitted after tests for bilinearity on the original MA(1) model are performed.

4.5.3 State dependent models

Using the concept of 'Volterra expansions', Priestley (1980) shows that a general relationship between x_t and ϵ_t can be represented as

$$x_t = \mathscr{F}(x_{t-1}, \ldots, x_{t-p}, \epsilon_{t-1}, \ldots, \epsilon_{t-q}) + \epsilon_t. \tag{4.26}$$

If $\mathscr{F}$ is assumed analytic, the right hand side of (4.26) can be expanded in a Taylor's series expansion about an arbitrary but fixed time point, allowing the relationship to be written as the State Dependent Model (SDM) of order (p,q)

$$x_t - \sum_{i=1}^{p} \phi_i(\mathbf{x}_{t-1})x_{t-i} = \mu(\mathbf{x}_{t-1}) + \epsilon_t - \sum_{i=1}^{q} \theta_i(\mathbf{x}_{t-1})\epsilon_{t-i}, \tag{4.27}$$

where $\mathbf{x}_t$ denotes the state vector

$$\mathbf{x}_t = (\epsilon_{t-q+1}, \ldots, \epsilon_t, x_{t-p+1}, \ldots, x_t).$$

As Priestley (1980, p. 54) remarks, this model 'has a natural and appealing interpretation as a locally linear ARMA model in which the evolution of the process at time $(t-1)$ is governed by a set of AR coefficients $\{\phi_i\}$, a set of MA coefficients $\{\theta_i\}$, and a local mean μ, all of which depend on the "state" of the process at time $(t-1)$'.

Indeed, if μ, $\{\phi_i\}$ and $\{\theta_i\}$ are all taken as constants, i.e., as independent of

$\mathbf{x}_{t-1}$, (4.27) reduces to the usual ARMA(p,q) model. Moreover, if only μ and $\{\phi_i\}$ are taken as constants but we set

$$\theta_i(\mathbf{x}_{t-i}) = \theta_i + \sum_{j=1}^{p} \delta_{ij} x_{t-j}, \quad i = 1, \ldots, q,$$

then the SDM reduces to the bilinear model (4.22), with the $\theta_i(\mathbf{x}_{t-i})$ being linear functions of $(x_{t-1}, \ldots, x_{t-p})$.

The SDM class of non-linear models can also be shown to include the *threshold AR* model (Tong and Lim, 1980), the *exponential AR* model (Haggan and Ozaki, 1981), and various other non-linear specifications that have been developed over recent years. Although Haggan, Heravi and Priestley (1984) provide an extensive study of the application of SDMs to a wide variety of non-linear time series, no financial, or even economic, data are used and it must remain to be seen whether such models are found to be useful for the modelling of financial time series.

4.5.4 *Asymmetric time series models*

It is common practice in much applied work to use some form of transformation to turn a skewed marginal distribution into a more normal distribution and, as we have seen, financial time series are often characterised by such distributions. Wecker (1981) argues, however, that observed skewness in a time series might result from the series being generated by an *asymmetric* model, i.e., one in which x_t responds in a different fashion to an innovation a_t depending on whether the innovation is positive or negative. This may be illustrated by the asymmetric first order moving average process defined as

$$x_t = a_t - \theta^+ a_{t-1}^+ - \theta^- a_{t-1}^-,$$

where

$$a_t^+ = \max(a_t, 0)$$

and

$$a_t^- = \min(a_t, 0)$$

are positive and negative innovations respectively. Wecker shows that, unlike the symmetric MA(1) process where the mean of x_t is zero, the mean of the asymmetric MA(1) process is

$$\mu = \frac{\theta^- - \theta^+}{\sqrt{2\pi}}$$

if we assume, for simplicity, that $a_t \sim NID(0,1)$. The asymmetric series x_t has variance

$$E(x_t-\mu)^2=\sigma^2=1+\frac{(\theta^+)^2+(\theta^-)^2}{2}-\mu^2,$$

and lag one autocovariance

$$E(x_t-\mu)(x_{t-1}-\mu)=\gamma_1=-\frac{\theta^+ + \theta^-}{2},$$

with $\gamma_k=0$ for $k\geq 2$. The symmetric MA(1) process is obtained when $\theta^+=\theta^-=\theta$, in which case the familiar results $\mu=0$, $\sigma^2=1+\theta^2$ and $\gamma_1=-\theta$ are obtained. Moreover, if $\theta^-=-\theta^+$, $\gamma_1=0$ and the asymmetric process becomes indistinguishable from white noise with mean $(-2\theta^+)/\sqrt{2\pi}$. Using this mean to make forecasts will result in a forecast error variance of

$$1+\left(\frac{\pi-2}{\pi}\right)(\theta^+)^2,$$

rather than the true innovation variance of one. For $-\theta^-=\theta^+=-0.9$, this represents an increase in forecast error variance of about 30 per cent over what would be possible if the forecasts were made using the true asymmetric model.

Note that the sample ACF function will give no clue as to whether the series is being generated by an asymmetric or a symmetric process, and any observed skewness in the marginal distribution of x_t may well prompt the application of a power transformation. Wecker suggests that, unknown to analysts, asymmetric time series may abound, but are analysed as either white-noise processes, if $\theta^- \simeq -\theta^+$, or as symmetric processes of power transformed series.

Wecker provides details of ML estimation and of a test for symmetry, and investigates the asymmetric MA(1) model on a variety of industrial price change series. Some of these series are quoted prices and, it is argued, should thus be asymmetric, while others, being transactions prices, should be symmetric. Wecker confirms these predictions, and also finds that for all asymmetric series the estimate of θ^- is large and positive, while that for θ^+ is small and non-positive, thus providing support for the view that when market conditions change, quoted prices are not revised immediately, this delay operating more strongly against reductions in price quotations than against increases.

4.6 The Hamilton two-state switching-regime model

Hamilton (1989) and Lam (1990) propose a switching-regime Markov model which can be regarded as a non-linear extension of an ARMA model that can accommodate complicated dynamics such as asymmetry and conditional heteroskedasticity. The set up is that of the UC model developed in chapter 3.2, i.e.

$$x_t = z_t + u_t, \tag{4.28}$$

where again z_t is a non-stationary random walk component, but where its drift now evolves according to a two-state Markov process

$$z_t = \mu(S_t) + z_{t-1} = \alpha_0 + \alpha_1 S_t + z_{t-1}, \tag{4.29}$$

where

$$P(S_t = 1 \mid S_{t-1} = 1) = p,$$
$$P(S_t = 0 \mid S_{t-1} = 1) = 1 - p,$$
$$P(S_t = 1 \mid S_{t-1} = 0) = 1 - q,$$
$$P(S_t = 0 \mid S_{t-1} = 0) = q.$$

The component u_t is assumed to follow an AR(r) process

$$\phi(B) u_t = \epsilon_t, \tag{4.30}$$

where the innovation sequence $\{\epsilon_t\}$ is strict white noise, but the case when $\phi(B)$ contains a unit root is allowed: unlike the conventional UC specification, u_t can be non-stationary. In fact, a special case of the conventional UC model results when $p = 1 - q$: the random walk component then has an innovation restricted to being a two-point random variable, taking the values 0 and 1 with probabilities q and $(1 - q)$ respectively, rather than a zero mean random variable drawn from a continuous distribution such as the normal.

The stochastic process for S_t is strictly stationary, having the AR(1) representation

$$S_t = (1 - q) + \lambda S_{t-1} + V_t,$$

where $\lambda = p + q - 1$ and where the innovation V_t has the conditional probability distribution

$$P(V_t = (1 - p) \mid S_{t-1} = 1) = p,$$
$$P(V_t = -p \mid S_{t-1} = 1) = 1 - p,$$
$$P(V_t = -(1 - q) \mid S_{t-1} = 0) = q,$$
$$P(V_t = q \mid S_{t-1} = 0) = 1 - q.$$

This innovation is uncorrelated with lagged values of S_t, for

$$E[V_t \mid S_{t-j} = 1] = E[V_t \mid S_{t-j} = 0] = 0 \text{ for } j \geq 1,$$

but it is not independent of such lagged values, since, for example,

$$E[V_t^2 \mid S_{t-1}=1]=p(1-p),$$
$$E[V_t^2 \mid S_{t-1}=0]=q(1-q).$$

The variance of the Markov process can be shown to be

$$\alpha_1^2 \frac{(1-p)(1-q)}{(2-p-q)^2}.$$

As this variance approaches zero, i.e., as p and q approach unity, so the random walk component (4.29) approaches a deterministic trend. If $\phi(B)$ contains no unit roots, x_t will thus approach a TS process, whereas if $\phi(B)$ does contain a unit root, x_t approaches a DS process.

Given $\{x_t\}_0^T$, ML estimates of the model are obtained by first expressing (4.28) as

$$u_t = u_{t-1} - x_t - x_{t-1} - \alpha_0 - \alpha_1 S_t,$$

and solving backwards in time to yield

$$u_t = x_t - x_0 - \alpha_0 t - \alpha_1 \sum_{i=1}^{t} S_i + u_0. \tag{4.31}$$

Using (4.30) and (4.31), the innovations ϵ_t can be expressed as

$$\epsilon_t = \phi(B)[x_t - x_0 - \alpha_0 t] + \phi(1)u_0 - \alpha_1\phi(1)\sum_{i=1}^{t} S_i + \alpha_1 \sum_{j=1}^{r}\left(\sum_{k=j}^{r}\phi_k\right)S_{t-j+1}.$$

Assuming that the innovations are normal, this expression can be utilised to calculate the log likelihood function on noting that this can be decomposed as the sum of the conditional (on past observations) log likelihoods. These conditional log likelihoods depend on unobserved current and past realisations of the Markov states. A recursive relationship can be shown to hold between the conditional distribution of the states and the conditional likelihood of the observations and this can be exploited to obtain an algorithm for evaluating the log likelihood function. Inferences about the unobserved components and states are then obtained as byproducts of this evaluation; details of the algorithm may be found in Hamilton (1989) and Lam (1990).

While this particular form of the model has yet to be fitted to financial time series, Pagan and Schwert (1990) consider a variant of it for modelling conditional stock volatility. Suppose that u_t contains a unit root and that its innovation is no longer strict white noise but is given by the product process $\epsilon_t = \sigma(S_t)U_t$, where $U_t \sim NID(0,1)$ and

$$\sigma(S_t) = \omega_0 + \omega_1 S_t,$$

with U_t independent of all S_{t-j}. The model can then be written as

$$\nabla x_t - \mu(S_t) = \sum_{i=1}^{r-1} \phi_i(\nabla x_{t-i} - \mu(S_{t-i})) + \sigma(S_t)U_t .$$

Within this set up, the two states can be interpreted as high ($S_t = 1$) and low ($S_t = 0$) variance regimes. On estimating the model using monthly US stock return data from July 1835 to December 1925, Pagan and Schwert found that, although the mean return was not a function of the regime (both $\hat{\alpha}_0$ and $\hat{\alpha}_1$ were insignificant), the conditional variance $\sigma(S_t)$ certainly was. The estimates of the Markov probabilities were $\hat{p} = 0.90$ and $\hat{q} = 0.96$, so that the high variance regime is less likely to occur than the low variance regime, but both regimes are likely to persist once they occur, the expected duration of the regimes being $(1-\hat{p})^{-1} = 10.4$ months and $(1-\hat{q})^{-1} = 26.2$ months.

4.7 Non-linear dynamics and chaos

The processes introduced so far in this chapter all have in common the aim of modelling *stochastic* non-linearities in financial time series. This would seem a natural approach to take by those used to dealing with stochastic time series processes, but recently a literature has developed that considers the question of whether such series could be generated, at least in part, by non-linear *deterministic* laws of motion. This has been prompted by findings in the natural sciences of completely deterministic processes that generate behaviour which looks random under standard statistical tests: processes that are termed 'deterministic chaos'. A simple example is provided by Brock (1986), where a formal development of deterministic chaos models is provided. Consider the difference equation

$$x_t = f(x_{t-1}),\ x_0 \in [0,1], \tag{4.32}$$

where

$$f(x) = \begin{cases} x/\alpha, & x \in [0,\alpha] \\ (1-x)/(1-\alpha), & x \in [\alpha,1],\ 0 < \alpha < 1 \end{cases} .$$

Most trajectories of this difference equation generate the same ACFs as an AR(1) process for x_t with parameter $\phi = (2\alpha - 1)$. Hence, for $\alpha = \frac{1}{2}$, the trajectory will be indistinguishable from white noise, although it has been generated by a purely deterministic non-linear process: for further discussion of this particular function, known as a *tent map* because the graph of x_t against x_{t-1} is shaped like a 'tent', see Hsieh (1991), who also considers other relevant examples of chaotic systems.

Are such models useful in finance? Brock (1988) considers some models of equilibrium asset pricing that might lead to chaos and complex dynamics: the idea that there should be no arbitrage profits in financial equilibrium is linked with the theory of economic growth to show how dynamics in the 'dividend' process are transmitted through the equilibrating mechanism to equilibrium asset prices. These dynamics can be linear, non-linear or chaotic depending on the constraints imposed on the models.

4.8 Testing for non-linearity

As the previous sections have demonstrated, there has been a wide variety of non-linear models proposed for modelling financial time series. We have, in particular, compared and contrasted the ARCH and bilinear models, and in so doing have discussed LM tests for each. Nevertheless, given the range of alternative non-linear models, it is not surprising that a number of other tests for non-linearity have also been proposed, but since the form of the departure from linearity is often difficult to specify a priori, many tests are 'diagnostic' in nature, i.e., a clear alternative to the null hypothesis of linearity is not specified, and this, of course, leads to difficulties in discriminating between the possible causes of 'non-linear misspecification'.

Keenan (1985) and Tsay (1986) provide a regression type test that appears to have good power against the non-linear moving average (see Robinson, 1977) and bilinear alternatives, but possibly has low power against ARCH models. This uses the Volterra expansion of x_t employed by Priestley (1980) to obtain (4.26), which allows x_t to be written in the very general form

$$x_t=\mu+\sum_{i=-\infty}^{\infty}\psi_i a_{t-i}+\sum_{i,j=-\infty}^{\infty}\psi_{ij}a_{t-i}a_{t-j}+\sum_{i,j,k=-\infty}^{\infty}\psi_{ijk}a_{t-i}a_{t-j}a_{t-k}+\dots \tag{4.33}$$

Obviously, x_t will be non-linear if any of the higher-order coefficients $\{\psi_{ij}\}$, $\{\psi_{ijk}\}, \dots$ are non-zero: if not, we obtain the usual linear Wold decomposition. This test of non-linearity is, then, essentially a test of no multiplicative terms in (4.33) and is based on a sequence of regressions using x_t and its cross-products. A modification of this test, based on examining unconditional third-order moments, has been suggested by Hsieh (1989b, 1991). Tests based on the estimated bispectrum have also been developed: see, for example, Ashley, Patterson and Hinich (1986) and Brockett, Hinich and Patterson (1988), although these too appear to have low power against ARCH alternatives.

A further test that has created considerable interest is the BDS statistic, based on the concept of the *correlation integral*: see, for example, Brock (1986), Brock and Sayers (1988), Brock and Dechert (1988), Brock and Baek (1991), Hsieh (1989b, 1991), Ramsey and Yuan (1990), Ramsey, Sayers and Rothman (1990), Scheinkman and LeBaron (1989) and

Scheinkman (1990). For an observed series $\{x_t\}_1^T$ the correlation integral $C_N(l,T)$ is defined as

$$C_N(l,T)=\frac{2}{T_N(T_N-1)}\sum_{t<s} I_l(x_t^N,x_s^N),$$

where

$$x_t^N=(x_t,x_{t+1},\ldots,x_{t+N-1})$$

and

$$x_s^N=(x_s,x_{s+1},\ldots,x_{s+N-1})$$

are called 'N-histories', $I_l(x_t^N,x_s^N)$ is an indicator function that equals one if $\|x_t^N-x_s^N\|<l$ and zero otherwise, $\|\cdot\|$ being the sup-norm, and $T_N=T-N+1$.

The correlation integral is an estimate of the probability that any two N-histories, x_t^N and x_s^N, are within l of each other. From it we may define the *correlation dimension* of $\{x_t\}$ as

$$\nu=\lim_{l\to 0}\,[\ln C_N(l,T)/\ln l],\ \text{if the limit exists.}$$

Physicists use the correlation dimension to distinguish between chaotic deterministic systems and stochastic systems, the intuition being that if the data is generated by a chaotic system then the dimension ν should be 'small'. Unfortunately, there is little formal statistical theory to underpin this intuition: indeed, the estimated correlation dimension may be substantially biased even in samples of 2,000: see Ramsey and Yuan (1990). An alternative strategy uses the result that if the x_ts are iid, then

$$C_N(l,T)\to C_1(l,T)^N,\ \text{as } T\to\infty$$

and

$$w_N(l,T)=\sqrt{T}[C_N(l,T)-C_1(l,T)^N]/\sigma_N(l,T)$$

has a standard normal limiting distribution, where the expression for the variance $\sigma_N^2(l,T)$ may be found in, for example, Hsieh (1989b, p. 343). Thus the BDS statistic $w_N(l,T)$ tests the null hypothesis that a series is iid: it is a diagnostic test since a rejection of this null is consistent with some type of dependence in the data, which could result from a linear stochastic system, a non-linear stochastic system, or a non-linear deterministic system. Additional diagnostic tests are therefore needed to determine the source of the rejection, but simulation experiments do suggest that the BDS test has power against simple linear deterministic systems as well as non-linear stochastic processes.

It should be emphasised, however, that all these tests are designed to distinguish between linear and non-linear *stochastic* dynamics: they are not, as yet, capable of distinguishing non-linear stochastic dynamics from deterministic chaotic dynamics, although the rejection of linearity may, of course, motivate the investigation of chaotic models.

Given that the presence of important non-linear dynamics may, for example, make standard linear rational expectations models quite misleading, what evidence has so far accumulated concerning the presence of non-linear dynamics? Apart from the applications of the specific non-linear models discussed earlier, there have been numerous recent studies providing general tests of non-linearity in financial time series. Ashley and Patterson (1986, 1989), Scheinkman and LeBaron (1989), Hinich and Patterson (1985, 1989), De Gooijer (1989), Ramsey (1990) and Hsieh (1991) all present evidence that stock returns contain important stochastic non-linearities, even though they use different tests on data collected over different sampling periods and intervals. Both Brockett, Hinich and Patterson (1988) and Hsieh (1989a, 1989b) find similar non-linear stochastic dependence in daily exchange rates, while Frank and Stengos (1989) examine rates of return on gold and silver.

Is there any evidence for chaotic dynamics? Although there are major difficulties in testing for chaos using the typically small (relative to the natural sciences) data sets found in finance and economics, Ramsey, Sayers and Rothman (1990) find no evidence whatsoever, from analysis of the correlation dimension, for such dynamic behaviour in the weekly stock return data analysed by Scheinkman and LeBaron (1989). Similarly, Hsieh (1991) finds no evidence of chaotic dynamics in an extended sample of weekly stock returns, although there is considerable evidence of ARCH-type non-linearity.

5 Regression techniques for non-integrated financial time series

The analysis of the general linear regression model forms the basis of every standard econometrics text and we see no need to repeat such a development here. Models relating financial time series, however, often cannot be analysed within the basic framework of ordinary least squares regression, or even its extensions incorporating generalised least squares or instrumental variables techniques. This chapter thus develops a general theory of regression, based on the work of Hansen (1982), White (1984) and White and Domowitz (1984), that builds upon the univariate time series techniques of the previous chapters and is applicable to many, but by no means all, of the regression problems that arise in the analysis of the relationships between financial time series.

Section 1 thus sets out the basic dynamic linear regression model, while section 2 incorporates ARCH error effects into the framework. Misspecification testing is the topic of section 3 and the multivariate linear regression model is briefly introduced in section 4. This paves the way for more general multivariate regression techniques and the remaining sections of the chapter deal with vector autoregressions and its various extensions, including a discussion of the concepts of exogeneity and causality.

5.1 Regression models

5.1.1 Regression with non-integrated time series

We now extend our modelling techniques to consider relationships between a group of time series $\{\mathbf{z}_t\}$. We begin by analysing the simplest case in which a *single* variable y_t is postulated to be a (linear) function of past values of itself and present and past values of a vector of other variables $\mathbf{x}_t = (x_{1t}, \ldots, x_{kt})$. Here $\mathbf{z}_t = (y_t, \mathbf{x}_t)$ and, for the observed realisation $\{\mathbf{z}_t\}_1^T$, the model can be written as

$$y_t = \alpha_0 + \sum_{i=1}^{m} \alpha_i y_{t-i} + \sum_{i=0}^{m} \beta_i \mathbf{x}_{t-i} + u_t, \quad m+1 \le t \le T, \tag{5.1}$$

or, in matrix form,

$$\mathbf{y} = \mathbf{X}\beta + \mathbf{u}, \tag{5.2}$$

where

$$\begin{aligned}
\mathbf{y} &= (y_{m+1}, \ldots, y_T)' \\
\mathbf{X} &= (\mathbf{X}_{m+1}, \ldots, \mathbf{X}_T)' \\
\mathbf{X}_t &= (1, y_{t-1}, \ldots, y_{t-m}, \mathbf{x}_t, \ldots, \mathbf{x}_{t-m}) \\
\mathbf{u} &= (u_{m+1}, \ldots, u_T)' \\
\beta &= (\alpha_0, \alpha_1, \ldots, \alpha_m, \beta_0, \ldots, \beta_m)' \\
\beta_i &= (\beta_{i1}, \ldots, \beta_{ik}).
\end{aligned}$$

To estimate the parameters of interest, contained in the vector β, certain assumptions are needed about $\{\mathbf{z}_t\}_1^T$ and the error process $\{u_t\}_1^T$. We begin by assuming that $\{\mathbf{z}_t\}_1^T$ is a normally distributed stationary stochastic process. Noting that $\mathbf{z}_t$ is $k+1$-dimensional, extension of the stationarity requirements for a univariate series given in chapter 2.1 to this multivariate setting yields

$$E(\mathbf{z}_t) = \boldsymbol{\mu} = (\mu_y, \mu_1, \ldots, \mu_k)$$

and

$$Cov(\mathbf{z}_t, \mathbf{z}_s) = \Gamma(|t-s|), \ 1 \le t, s \le T,$$

so that the $\mathbf{z}_t$s have identical means and variances and their temporal covariances depend only on the absolute value of the time between them. Note, however, that the assumption of stationarity alone is not sufficient to obtain an operational model of the form (5.1). This is because the non-zero covariances allow dependence between, for example, $\mathbf{z}_1$ and $\mathbf{z}_T$, implying that the lag length m in (5.1) should strictly be set at $t-1$, so that the number of unknown parameters in the model increases with T. We thus need to restrict the form of dependence in $\{\mathbf{z}_t\}_1^T$ and, to this end, the following concepts are important (see White, 1984, and Spanos, 1986, for detailed formal discussion): $\mathbf{z}_t$ is said to be *asymptotically independent* if

$$\Gamma(\tau) \to 0 \text{ as } \tau = |t-s| \to \infty,$$

and *ergodic* if

$$\lim_{t\to\infty} \left(\frac{1}{T} \sum_{\tau=1}^{T} \Gamma(\tau) \right) = 0.$$

It is conventional to make either the assumption of asymptotic independence or the somewhat weaker assumption of ergodicity (cf. the univariate development in chapter 2), and this allows us to restrict the *memory* of the process $\{\mathbf{z}_t\}_1^T$ and hence to fix the maximum lag at an appropriate value, say m, in (5.1). We will also assume that the error $\{u_t\}_1^T$, defined formally as

$$u_t = y_t - E(y_t \mid \mathbf{y}_{t-1}^0, \mathbf{x}_t^0),$$

where

$$\mathbf{y}_{t-1}^0 = (y_{t-1}, y_{t-2}, \ldots, y_1)$$

and

$$\mathbf{x}_t^0 = (\mathbf{x}_t, \mathbf{x}_{t-1}, \ldots, \mathbf{x}_1),$$

satisfies the following properties

$$E(u_t) = E(u_t \mid \mathbf{y}_{t-1}^0, \mathbf{x}_t^0) = 0$$

$$E(u_t u_s) = E\{E(u_t u_s \mid \mathbf{y}_{t-1}^0, \mathbf{x}_t^0)\} = \begin{cases} \sigma^2 & t=s \\ 0 & t>s \end{cases}.$$

These two properties define u_t to be a martingale difference relative to the 'history' $(\mathbf{y}_{t-1}^0, \mathbf{x}_t^0)$ having bounded variance, i.e., it is an *innovation* process. Note that the assumption of asymptotic independence implies that the roots of the polynomial $\mathscr{Z}^m - \sum_1^m a_i \mathscr{Z}^{m-i} = 0$ are all less than unity in absolute value.

Assuming $\mathbf{X}$ to be of full rank $K = (m+1)(k+1)$, so that $\mathbf{X}'\mathbf{X}$ is non-singular, and u_t to be $NID(0, \sigma^2)$, the LS (and approximate ML) estimator of β obtained using the sample $\{\mathbf{z}_t\}_1^T$ is

$$\hat{\beta}_T = (\mathbf{X}'\mathbf{X})^{-1}\mathbf{X}'\mathbf{y},$$

while the LS and approximate ML estimators of σ^2 are

$$\hat{\sigma}_T^2 = (T-m)^{-1}\hat{\mathbf{u}}'\hat{\mathbf{u}} \text{ and } \tilde{\sigma}_T^2 = T^{-1}\hat{\mathbf{u}}'\hat{\mathbf{u}}$$

respectively, where $\hat{\mathbf{u}} = \mathbf{y} - \mathbf{X}\hat{\beta}_T$ (the ML estimators are said to be approximate because the initial conditions involving the observations $y_1, \ldots, y_m$ are ignored). Since u_t is not independent of future y_ts, $E(\mathbf{X}'\mathbf{u}) \neq 0$, and so $\hat{\beta}_T$ is a biased estimator of β

$$E(\hat{\beta}_T - \beta) = E[(\mathbf{X}'\mathbf{X})^{-1}\mathbf{X}'\mathbf{u}] \neq 0.$$

However, assuming $\mathbf{G}_T = E(\mathbf{X}'\mathbf{X}/T)$ to be uniformally positive definite and since $E(\mathbf{X}_t' u_t) = 0$ then, under certain conditions concerning the magnitude of $E(\mathbf{X}'\mathbf{X})$, $\hat{\beta}_T$ can be shown to be a *strongly consistent* estimator of β, as indeed is $\hat{\sigma}_T^2$ of σ^2. The estimators are also *asymptotically normal*

$$\mathbf{G}_T^{-\frac{1}{2}} T^{\frac{1}{2}}(\hat{\beta}_T - \beta) \overset{a}{\sim} N(\mathbf{0}, \mathbf{I})$$
$$T^{\frac{1}{2}}(\hat{\sigma}_T^2 - \sigma^2) \overset{a}{\sim} N(\mathbf{0}, 2\sigma^4),$$

(for formal derivations of these results, see, for example, White, 1984, and Spanos, 1986). $\mathbf{G}_T$ can be consistently estimated in this case as $\hat{\sigma}_T^2(\mathbf{X}'\mathbf{X})^{-1}$, this being the conventional covariance formula in LS regression.

These results can be extended to allow both $\mathbf{z}_t$ and u_t to exhibit time dependence and heterogeneity simultaneously. Specifically, the memory requirement can be relaxed from that of stationarity and asymptotic independence (or ergodicity) to one of *mixing* although, by way of tradeoff, certain moment restrictions have to be strengthened. *Strong mixing* and *uniform mixing* are generalisations of the concept of asymptotic independence that provide absolute and relative measures of temporal dependence, respectively, and also allow processes to exhibit considerable heterogeneity. White (1984, exercise 3.51 and theorem 4.25) provides a formal statement of the required conditions and shows that, in these circumstances, $\hat{\beta}_T$ is still consistent and asymptotically normal, although we now have

$$\mathbf{D}_T^{-\frac{1}{2}} T^{\frac{1}{2}}(\hat{\beta}_T - \beta) \overset{a}{\sim} N(\mathbf{0}, \mathbf{I}),$$

where

$$\mathbf{D}_T = (\mathbf{X}'\mathbf{X}/T)^{-1}\hat{\mathbf{V}}_T(\mathbf{X}'\mathbf{X}/T)^{-1}.$$

$\hat{\mathbf{V}}_T$ is an estimate of $\mathbf{V}_T = E(\mathbf{X}'\mathbf{uu}'\mathbf{X}/T)$, which can be expressed in terms of individual observations as

$$\mathbf{V}_T = E\left(T^{-1}\sum_{t=1}^{T}\mathbf{X}_t'u_tu_t'\mathbf{X}_t\right) = T^{-1}\sum_{t=1}^{T}E(\mathbf{X}_t'u_tu_t'\mathbf{X}_t)$$
$$+ T^{-1}\sum_{\tau=1}^{T-1}\sum_{t=\tau+1}^{T}E(\mathbf{X}_t'u_tu_{t-\tau}'\mathbf{X}_{t-\tau} + \mathbf{X}_{t-\tau}'u_{t-\tau}u_{t-\tau}'\mathbf{X}_t)$$
$$= T^{-1}\sum_{t=1}^{T}V(\mathbf{X}_t'u_t)$$
$$+ T^{-1}\sum_{\tau=1}^{T-1}\sum_{t=\tau+1}^{T}(Cov(\mathbf{X}_t'u_t, \mathbf{X}_{t-\tau}'u_{t-\tau}) + Cov(\mathbf{X}_{t-\tau}'u_{t-\tau}, \mathbf{X}_t'u_t)),$$

thus revealing that $\mathbf{V}_T$ is the average of the variances of $\mathbf{X}_t'u_t$ plus a term that takes into account the covariances between $\mathbf{X}_t'u_t$ and $X_{t-\tau}'u_{t-\tau}$ for all t and τ. With our mixing assumptions, the covariance between $\mathbf{X}_t'u_t$ and $\mathbf{X}_{t-\tau}'u_{t-\tau}$ goes to zero as $\tau \to \infty$, and hence $\mathbf{V}_T$ can be approximated by

$$\tilde{\mathbf{V}}_T = T^{-1}\sum_{t=1}^{T}E(\mathbf{X}_t'u_tu_t'\mathbf{X}_t) \tag{5.3}$$

$$+T^{-1}\sum_{\tau=1}^{n}\sum_{t=\tau+1}^{T} E(\mathbf{X}_t'u_tu_{t-\tau}'\mathbf{X}_{t-\tau}+\mathbf{X}_{t-\tau}'u_{t-\tau}u_{t-\tau}'\mathbf{X}_t)$$

for some value n, because the neglected terms (those with $n<\tau\leq T$) will be small in absolute value if n is sufficiently large. Note, however, that if n is simply kept fixed as T grows, then the number of neglected terms grows, and may grow in such a way that the sum of these terms does not remain negligible. The estimator $\hat{\mathbf{V}}_T$, obtained by replacing u_t with $\hat{u}_t$ in (5.3), will then be a consistent estimator of $\tilde{\mathbf{V}}_T$ (and hence of $\mathbf{V}_T$) if n does not grow too fast as T grows: specifically, we must ensure that n grows more slowly than $T^{\frac{1}{3}}$.

5.1.2 Hypothesis testing

As is traditional, we consider hypotheses that can be expressed as linear combinations of the parameters in β

$$\mathbf{R}\beta=\mathbf{r},$$

where $\mathbf{R}$ and $\mathbf{r}$ are a matrix and a vector of known elements, both of row dimension q, that specify the q hypotheses of interest.

Several different approaches can be taken in computing a statistic to test the null hypothesis $\mathbf{R}\beta=\mathbf{r}$ against the alternative $\mathbf{R}\beta\neq\mathbf{r}$: we will consider here the use of Wald, Lagrange multiplier and (quasi-)likelihood ratio statistics. Although the approaches to forming the test statistics differ, in each case an underlying asymptotic normality property is exploited to obtain a statistic which is distributed asymptotically as χ^2. Detailed development of the theory of hypothesis testing using these approaches may be found in Godfrey (1988).

The Wald statistic allows the simplest analysis, although it may not be the easiest to compute. Its motivation is the observation that, when the null hypothesis is correct, $\mathbf{R}\hat{\beta}_T$ should be close to $\mathbf{R}\beta=\mathbf{r}$, so that a value of $\mathbf{R}\hat{\beta}_T-\mathbf{r}$ far from zero should be viewed as evidence against the null hypothesis. To tell how far from zero $\mathbf{R}\hat{\beta}_T-\mathbf{r}$ must be before we reject the null hypothesis, we need to determine its asymptotic distribution. White (1984) shows that, if the rank of $\mathbf{R}$ is $q\leq K$, then the Wald statistic is

$$\mathcal{W}_T=T(\mathbf{R}\hat{\beta}_T-\mathbf{r})'\hat{\Omega}_T^{-1}(\mathbf{R}\hat{\beta}_T-\mathbf{r})\overset{a}{\sim}\chi_q^2, \tag{5.4}$$

where

$$\hat{\Omega}_T=\mathbf{R}\mathbf{D}_T\mathbf{R}'=\mathbf{R}(\mathbf{X}'\mathbf{X}/T)^{-1}\hat{\mathbf{V}}_T(\mathbf{X}'\mathbf{X}/T)^{-1}\mathbf{R}'.$$

This version of the Wald statistic is useful regardless of the presence of heteroskedasticity or serial correlation in the error $\mathbf{u}$ because a consistent estimator $\hat{\mathbf{V}}_T$ is used to construct $\hat{\Omega}_T$. In the special case when $\mathbf{u}$ is white noise, $\hat{\mathbf{V}}_T$ can be consistently estimated by $\hat{\sigma}_T^2(\mathbf{X}'\mathbf{X}/T)$, and the Wald statistic then has the form

$$\mathscr{W}_T = T(\mathbf{R}\hat{\beta}_T - \mathbf{r})'[\mathbf{R}(\mathbf{X}'\mathbf{X}/T)^{-1}\mathbf{R}']^{-1}(\mathbf{R}\hat{\beta}_T - \mathbf{r})/\hat{\sigma}_T^2,$$

which is simply q times the standard F-statistic for testing the hypothesis $\mathbf{R}\beta = \mathbf{r}$. The validity of the asymptotic χ_q^2 distribution for this statistic, however, depends crucially on the consistency of the estimator $\hat{\sigma}_T^2(\mathbf{X}'\mathbf{X}/T)$ for $\mathbf{V}_T$: if this $\hat{\mathbf{V}}_T$ is not consistent for $\mathbf{V}_T$, the asymptotic distribution of this form for $\mathscr{W}_T$ is not χ_q^2 and hence failure to take account of serial correlation and heterogeneity in the errors will lead to inferences being made using an incorrect distribution.

The Wald statistic is the most convenient test to use when the restrictions $\mathbf{R}\beta = \mathbf{r}$ are not easy to impose in estimating β. When these restrictions can be easily imposed, the Lagrange Multiplier statistic is more convenient to compute. The motivation for the LM statistic is that a constrained LS estimator can be obtained by solving the first-order conditions of the Lagrangian expression

$$\mathscr{L} = (\mathbf{y} - \mathbf{X}\beta)'(\mathbf{y} - \mathbf{X}\beta)/T + (\mathbf{R}\beta - \mathbf{r})'\lambda.$$

The Lagrange multipliers λ give the shadow price of the constraint and should therefore be small when the constraint is valid and large otherwise. The LM test can thus be thought of as testing the hypothesis that $\lambda = 0$. Solving the first-order conditions for λ yields

$$\ddot{\lambda}_T = 2(\mathbf{R}(\mathbf{X}'\mathbf{X}/T)^{-1}\mathbf{R}')^{-1}(\mathbf{R}\hat{\beta}_T - \mathbf{r}),$$

so that $\ddot{\lambda}_T$ is simply a non-singular transformation of $\mathbf{R}\hat{\beta}_T - \mathbf{r}$. Also provided by solving the first-order conditions is the constrained LS estimator $\ddot{\beta}_T$, given by

$$\ddot{\beta}_T = \hat{\beta}_T - (\mathbf{X}'\mathbf{X}/T)^{-1}\mathbf{R}'\ddot{\lambda}_T/2,$$

from which can be calculated the constrained estimator of σ^2

$$\ddot{\sigma}_T^2 = (T-m)^{-1}\ddot{\mathbf{u}}'\ddot{\mathbf{u}},$$

where $\ddot{\mathbf{u}} = \mathbf{y} - \mathbf{X}\ddot{\beta}_T$ are the residuals from the constrained regression. The LM test statistic is then defined as

$$\mathscr{L}\mathscr{M}_T = T\ddot{\lambda}_T'\Lambda_T^{-1}\ddot{\lambda}_T \overset{a}{\sim} \chi_q^2, \tag{5.5}$$

where

$$\Lambda_T = 4(\mathbf{R}(\mathbf{X}'\mathbf{X}/T)^{-1}\mathbf{R}')^{-1}\mathbf{R}(\mathbf{X}'\mathbf{X}/T)^{-1}\ddot{\mathbf{V}}_T(\mathbf{X}'\mathbf{X}/T)^{-1}\mathbf{R}'(\mathbf{R}(\mathbf{X}'\mathbf{X}/T)^{-1}\mathbf{R}')^{-1},$$

$\ddot{\mathbf{V}}_T$ being computed from the constrained regression. Note that the Wald and LM statistics (5.4) and (5.5) would be identical if $\hat{\mathbf{V}}_T$ were used in place of $\ddot{\mathbf{V}}_T$ and, indeed, the two statistics are asymptotically equivalent.

As we have seen, when the errors u_t are $NID(0,\sigma^2)$, the LS estimator $\hat{\beta}_T$ is also the ML estimator. When this is not the case, $\hat{\beta}_T$ is said to be a quasi-maximum likelihood (QML) estimator.

When $\hat{\beta}_T$ is the ML estimator, hypothesis tests can be based on the *log likelihood ratio* (LR)

$$\mathscr{LR}_T = \ln[\mathscr{L}(\ddot{\beta}_T,\ddot{\sigma}_T;\mathbf{y})/\mathscr{L}(\hat{\beta}_T,\hat{\sigma}_T;\mathbf{y})],$$

where

$$\mathscr{L}(\beta,\sigma;\mathbf{y}) = \exp\left[-T\ln\sqrt{2\pi} - T\ln\sigma - \tfrac{1}{2}\sum_{t=m+1}^{T}(y_t - \mathbf{X}_t\beta)^2/\sigma^2\right]$$

is the sample likelihood based on the normality assumption. Simple algebra yields the following alternative form of the statistic,

$$\mathscr{LR}_T = (T/2)\ln(\hat{\sigma}_T^2/\ddot{\sigma}_T^2),$$

and it can be shown that $-2\mathscr{LR}_T$ is asymptotically equivalent to the Wald statistic (5.4) and thus has the χ_q^2 distribution asymptotically, *provided* that $\hat{\sigma}_T^2(\mathbf{X}'\mathbf{X}/T)$ is a consistent estimator of $\mathbf{V}_T$. If this is not true, then $-2\mathscr{LR}_T$ is not asymptotically χ_q^2.

So far we have considered linear hypotheses of the form $\mathbf{R}\beta = \mathbf{r}$. In general, non-linear hypotheses can be conveniently represented as

$$H_0\colon \mathbf{s}(\beta) = \mathbf{0},$$

where $\mathbf{s}$ is a continuously differentiable function of β. Just as with linear restrictions, we can construct a Wald test based on the asymptotic distribution of $\mathbf{s}(\hat{\beta}_T)$, we can construct an LM test, or we can form a log likelihood ratio. Assuming that the rank of $\Delta\mathbf{s}(\beta) = q \le K$, where $\Delta\mathbf{s}$ is the gradient (derivative) of $\mathbf{s}$, then under H_0:$\mathbf{s}(\beta) = \mathbf{0}$, the Wald and LM test statistics are given by equations (5.4) and (5.5) with $\mathbf{s}(\hat{\beta}_T)$ and $\Delta\mathbf{s}(\hat{\beta}_T)$ replacing $\mathbf{R}\hat{\beta}_T - \mathbf{r}$ and $\mathbf{R}$, respectively, in (5.4), and $\mathbf{s}(\ddot{\beta}_T)$ and $\Delta\mathbf{s}(\ddot{\beta}_T)$ similarly replacing $\mathbf{R}\hat{\beta}_T - \mathbf{r}$ and $\mathbf{R}$ in (5.5).

5.1.3 Instrumental variable estimation

We have so far considered only (ordinary) LS estimation of the model (5.1). If the assumption $E(\mathbf{X}_t'u_t) = 0$ does not hold, but a set of l instrumental variables (IV), say $\mathbf{W}_t = (w_{1t}, \ldots, w_{lt})$, are available such that $E(\mathbf{W}_t'u_t) = \mathbf{0}$ and $E(\mathbf{W}'\mathbf{X}/T)$ has uniformly full column rank, then we can form the IV estimator

$$\tilde{\beta}_T = (\mathbf{X}'\mathbf{W}\hat{\mathbf{P}}_T\mathbf{W}'\mathbf{X})^{-1}\mathbf{X}'\mathbf{W}\hat{\mathbf{P}}_T\mathbf{W}'\mathbf{y},$$

where $\mathbf{W}=(\mathbf{W}_{m+1},\ldots,\mathbf{W}_T)$ and $\hat{\mathbf{P}}_T$ is a symmetric $l\times l$ positive definite norming matrix. For example, with $\mathbf{W}=\mathbf{X}$ and $\hat{\mathbf{P}}_T=(\mathbf{W}'\mathbf{W}/T)^{-1}$, $\tilde{\beta}_T=\hat{\beta}_T$, while for any $\mathbf{W}$, choosing $\hat{\mathbf{P}}_T=(\mathbf{W}'\mathbf{W}/T)^{-1}$ yields the *two-stage least squares* estimator. Analogous to the results for the LS estimator, if $\mathbf{W}_t$ is also mixing, then $\tilde{\beta}_T$ is strongly consistent and

$$\mathbf{D}_T^{-\frac{1}{2}}T^{\frac{1}{2}}(\tilde{\beta}_T-\beta)\overset{a}{\sim}N(\mathbf{0},\mathbf{I}),$$

where now

$$\mathbf{D}_T=(\mathbf{X}'\mathbf{W}\hat{\mathbf{P}}_T\mathbf{W}'\mathbf{X}/T^2)^{-1}(\mathbf{X}'\mathbf{W}/T)\hat{\mathbf{P}}_T\hat{\mathbf{V}}_T\hat{\mathbf{P}}_T(\mathbf{W}'\mathbf{X}/T)(\mathbf{X}'\mathbf{W}\hat{\mathbf{P}}_T\mathbf{W}'\mathbf{X}/T^2)^{-1}.$$

So far we have let $\hat{\mathbf{P}}_T$ be any positive definite matrix. By choosing $\hat{\mathbf{P}}_T=\hat{\mathbf{V}}_T^{-1}$, however, an *asymptotically efficient* estimator is obtained for the class of IV estimators with given instrumental variables $\mathbf{W}$, i.e.

$$\beta_T^*=(\mathbf{X}'\mathbf{W}\hat{\mathbf{V}}_T^{-1}\mathbf{W}'\mathbf{X})^{-1}\mathbf{X}'\mathbf{W}\hat{\mathbf{V}}_T^{-1}\mathbf{W}'\mathbf{y}$$

is asymptotically efficient within the class of IV estimators $\tilde{\beta}_T$.

How should we choose the set of instruments $\mathbf{W}_t$? It can be shown that the asymptotic precision of the IV estimator cannot be worsened by including additional instruments. There are situations, however, when nothing is gained by adding an extra instrument: this is when the additional instrument is uncorrelated with the residuals of the regression of $\mathbf{X}$ on the already included instruments.

When serial correlation or heteroskedasticity of unknown form is present in (5.1), there may, in fact, be no limit to the number of instrumental variables available for improving the efficiency of the IV estimator: functions of $\mathbf{X}$ and $\mathbf{W}$ are possible instruments. In the absence of serial correlation or heteroskedasticity, however, it is possible to specify precisely a finite set of instruments that yield the greatest possible efficiency: they will be those functions of $\mathbf{W}_t$ that appear in the conditional expectation of $\mathbf{X}_t$ given $\mathbf{W}_t$.

Example 5.1: Forward exchange rates as optimal predictors of future spot rates

An important illustration of these estimation techniques is found in the analysis of foreign exchange markets, where the efficient markets hypothesis becomes the proposition that the expected rate of return to speculation in the forward market, conditioned on available information, is zero. Hansen and Hodrick (1980) test this 'simple' efficiency hypothesis in the following way. Let s_t and $f_{t,k}$ be the logarithms of the spot exchange rate and the k-period forward rate determined at time t, respectively. Since $s_{t+k}-f_{t,k}$ is an approximate measure of the rate of return to speculation, the simple efficient markets hypothesis is that

$$f_{t,k}=E(s_{t+k}\,|\,\Phi_t),$$

where Φ_t is the information set available at time t. This implies that the speculative rate of return $y_{t+k}=s_{t+k}-f_{t,k}$ should be uncorrelated with information available at time t: for example, in the regression of the return on a constant and two lagged returns,

$$y_{t+k}=\alpha_0+\alpha_1 y_t+\alpha_2 y_{t-1}+u_{t+k},$$

the α_i, $i=0,1,2$, should all be zero. Assuming that y_t is mixing and that $E(y_{t-j}u_{t+k})=0$, for $j\geq 0$, which is easily verified, LS estimation provides consistent estimates of the α's. However, in the present circumstances, the forecast error $u_{t+k}=y_{t+k}-E(y_{t+k}\mid\Phi_t)$ will be serially correlated, so that the usual estimated covariance matrix will be inconsistent.

This serial correlation arises from the fact that the realised values of the spot exchange rate $s_{t+1},s_{t+2},\ldots,s_{t+k}$ are not known when the forward rate $f_{t,k}$ is set at time t, hence the corresponding k-period ahead forecast errors $u_{t+k-j}=s_{t+k-j}-f_{t-j,k}$, $j=1,2,\ldots,k-1$, are not observable. Since u_{t+1}, $u_{t+2},\ldots,u_{t+k-1}$ are not part of the available information set, we cannot rule out the possibility that $E(u_{t+k}\mid u_{t+k-j})\neq 0$, $1\leq j\leq k-1$, or that

$$Cov(u_{t+k},u_{t+k-j})\neq 0, \quad j=1,2,\ldots,k-1.$$

On the other hand, the preceding k-period forecast errors u_{t+k-j} for $j\geq k$ *are* observable. Efficiency thus requires $E(u_{t+k}\mid u_{t+k-j})=0$, $j\geq k$, and hence

$$Cov(u_{t+k},u_{t+k-j})=0, \quad j\geq k.$$

With our mixing assumptions concerning s_t and $f_{t,k}$, $u_{t+k}=s_{t+k}-f_{t,k}$ will also be mixing, and combining the above covariances shows that the forecast errors can be thought of as being generated by an MA($k-1$) process.

Can we use generalised least squares procedures to make inferences about the α's? The answer is no, because such techniques require the regressors to be *strictly exogenous*, which means that $E(u_{t+k}\mid\ldots,y_{t-1},y_t,y_{t+1},\ldots)=0$, i.e., that future y_t's would be useless in determining the optimal forecast for y_{t+k} (strict, and other forms of, exogeneity are formally discussed in section 5.5). This is clearly inappropriate since such values would provide useful information for forecasting future rates of return. The use of regressors that are not strongly exogenous renders generalised LS techniques inconsistent, because the transformation used to eliminate the serial correlation in the residuals makes the transformed residuals for some particular period linear combinations of the original residuals and their lagged values. These in turn are likely to be correlated with the transformed data for the same period, since these include current values of the variables in the information set.

One way of avoiding these difficulties is to choose the sampling interval to equal the forecast interval, i.e., to set $k=1$, in which case the forecast errors will be serially uncorrelated. This procedure of using *non-overlapping* data clearly does not make use of all the available information: $T(1-k^{-1})$ observations are sacrificed. In the present application weekly observations are typically used with k set at 13 (3-month forward exchange rates readily being available). Using non-overlapping data, i.e., sampling only every thirteen weeks, would thus throw away over 90 per cent of the available observations.

The complete data set can be used if we adjust the covariance matrix of $\hat{\beta}=(\hat{\alpha}_0,\hat{\alpha}_1,\hat{\alpha}_2)'$ in the appropriate fashion. As we have shown, a consistent covariance matrix is

$$\mathbf{D}_T=(\mathbf{X}'\mathbf{X}/T)^{-1}\hat{\mathbf{V}}_T(\mathbf{X}'\mathbf{X}/T)^{-1},$$

where now the columns making up the $\mathbf{X}$ matrix contain a constant and the two lagged values of y_{t+k}. An expression for $\hat{\mathbf{V}}_T$ is given by $\hat{\mathbf{V}}_T=T^{-1}\mathbf{X}'\hat{\Theta}\mathbf{X}$ and, from the fact that the residuals $\hat{u}_{t+k}$ follow an MA$(k-1)$ process, the elements of the $T\times T$ symmetric matrix $\hat{\Theta}$ will have the form

$$\hat{\Theta}_{i,i+j}=R(j),\quad i=1,2,\ldots,T-k+1,\quad j=0,1,\ldots,k-1$$
$$\hat{\Theta}_{i+j,i}=\hat{\Theta}_{i,i+j},$$

where

$$R(j)=T^{-1}\sum_{t=j+1}^{T}\hat{u}_{t+k}\hat{u}_{t+k-j},$$

and $\hat{\Theta}_{i,j}=0$ otherwise, i.e., $\hat{\Theta}$ is 'band diagonal', the band width being $2k-1$.

The hypothesis of market efficiency is $\beta=0$ and in the framework of section 5.1.2, $\mathbf{R}=\mathbf{I}_3$, $\mathbf{r}=\mathbf{0}$ and $\hat{\Omega}_T=\mathbf{D}_T$. The Wald statistic, for example, for testing this hypothesis takes the form

$$\mathscr{W}_T=T\hat{\beta}_T'\mathbf{D}_T^{-1}\hat{\beta}_T\overset{a}{\sim}\chi_3^2.$$

Hansen and Hodrick (1980) estimate regressions of this type for weekly data on spot and three-month ($k=13$) forward exchange rates for seven currencies (expressed in US cents per unit of foreign currency) from March 1973 to January 1979, and for three currencies relative to the UK pound sterling for certain episodes after the First World War, in this case using one-month ($k=4$) forward rates. Their findings indicate that the simple efficiency hypothesis is 'suspect' in both periods, but they offer a variety of reasons why this may be so, emphasising that rejection of the hypothesis $\beta=0$ cannot necessarily be identified with inefficiency in the foreign exchange market since certain intertemporal asset allocation and risk considerations are ignored in this formulation of the efficient markets hypothesis.

5.2 ARCH-in-mean regression models

5.2.1 The GARCH-M model

The estimation techniques developed above are applicable when little is known about the structure of the serial correlation and heteroskedasticity present in the errors in model (5.1). On certain occasions, however, it may be possible to specify the form of these departures from white noise, and a specification that has proved to be particularly useful in financial applications is the *(G)ARCH-in-Mean* [(G)ARCH-M] model proposed by Engle, Lilien and Robbins (1987) and employed by, for example, Domowitz and Hakkio (1985) for examining risk premia in the foreign exchange market and by French, Schwert and Stambaugh (1987) to model stock return volatility. Bollerslev, Chou and Kroner (1992) provide many further references to GARCH-M applications in finance, these often being attempts to model the linear relationship between the return and variance of a portfolio that emerges as a consequence of the intertemporal CAPM of Merton (1973, 1980).

The GARCH-M model extends the GARCH model developed in chapter 4.4 to the regression framework of equation (5.1)

$$y_t = \alpha_0 + \sum_{i=1}^{m} \alpha_i y_{t-i} + \sum_{i=0}^{m} \beta_i \mathbf{x}_{t-i} + \delta h_t^{\lambda} + u_t, \tag{5.6}$$

$$u_t = \epsilon_t - \sum_{i=1}^{r} \theta_i \epsilon_{t-i}, \tag{5.7}$$

$$E(\epsilon_t^2 \mid \Phi_{t-1}) = h_t = \gamma_0 + \sum_{i=1}^{p} \gamma_i \epsilon_{t-i}^2 + \sum_{i=1}^{q} \phi_i h_{t-i} + \vartheta \xi_t. \tag{5.8}$$

Here we allow the serially correlated errors u_t to be modelled as an MA(r) process (equation (5.7)), and the conditional variance h_t (conditional upon the information set at time $t-1$, Φ_{t-1}) both enters the 'mean' equation (5.6) and depends itself (equation (5.8)) upon a vector of explanatory variables ξ_t. Typically, λ is set at unity or a half, so that either the conditional variance or standard deviation is included in the mean equation. Under the assumption that the ϵ_t are $NID(0,\sigma^2)$, ML estimates of the GARCH-M model given by equations (5.6)–(5.8) can be obtained by maximising the likelihood function using the BHHH algorithm in a manner analogous to that discussed in chapter 4.4.2. There are some complications, however: for example, the information matrix is no longer block diagonal, so that all parameters must be estimated simultaneously, unlike the case discussed previously where the block diagonality of the information matrix allowed estimates of the parameters of the mean and conditional variance equations to be obtained by separate iterations.

If it is preferred, the alternative assumption that the ϵ_t follow a standardised t-distribution may be employed to allow more adequate modelling of the fat tails often found in the observed unconditional distributions of financial time series: Baillie and Bollerslev (1989), for example, provide the relevant expression for the log-likelihood function.

Example 5.2: Stock returns and volatility

An example of the GARCH-M model is provided by Baillie and DeGennaro (1990), who analyse the volatility of daily US stock returns using the following model:

$$y_t = \alpha + \beta x_t + \delta h_t^{\lambda} + u_t$$

$$u_t = \epsilon_t - \theta \epsilon_{t-1}$$

$$h_t = \gamma_0 + \gamma_1 \epsilon_{t-1}^2 + \phi h_{t-1} + \vartheta \xi_t .$$

In this model, y_t is the actual return on day t, x_t is a variable measuring the compensation received by the sellers of stock for payment delays caused by settlement and cheque-clearing procedures, this being proxied by the change from day $t-1$ to day t in the federal funds rate continuously compounded over the delay period, and ξ_t is the absolute value of this change, i.e., $\xi_t = |x_t|$. The inclusion of h_t^{λ} in the returns equation is an attempt to incorporate a measure of risk into the returns generating process and is an implication of the 'mean-variance hypothesis' underlying many theoretical asset pricing models, such as the intertemporal CAPM discussed above. The error u_t is assumed to be MA(1) to capture the effect of non-synchronous trading, while the innovations ϵ_t can be assumed to follow a conditional t distribution. The conditional variance equation is GARCH(1,1) with the addition of the variable ξ_t.

With risk-averse investors, α is expected to be positive, as is β, since compensation should increase as the payment delay period increases. Under the mean-variance hypothesis, $\delta > 0$, so that large values for the conditional variance are expected to be associated with large returns.

Baillie and DeGennaro estimate this model, and a number of variants of it, using daily returns data from 1 January 1970 to 22 December 1987, a total of 4,542 observations. Setting λ at either 1 or $\frac{1}{2}$ produces δ estimates that are positive but insignificant, as indeed are the estimates of α. Estimates of β, however, are significantly positive. Either the constant or ξ_t was included in the conditional variance equation, and were found to be significantly positive, while the estimates of $\gamma + \phi$ were found to be close to unity, indicating a high degree of persistence in h_t. The moving average innovation parameter θ was found to be highly significant, taking a value of about 0.25.

Baillie and DeGennaro conclude that any relationship between mean returns and return variance or standard deviation is weak, and that the results suggest that investors consider some other risk measure to be more important than the variance of portfolio returns: the traditional two-parameter asset pricing models relating portfolio means to variances would thus appear to be inappropriate.

Similar models are estimated by French, Schwert and Stambaugh (1987) for daily returns over a sample period running from January 1928 to December 1984, the total number of observations being in excess of 15,000. In their study, the variable y_t is the daily excess return, defined to be the market return minus the risk-free interest rate, and neither x_t nor ξ_t are included in the specification. Their estimate of δ is significantly positive, with a value for the exponent λ of a half being slightly preferred. A GARCH(1,2) model is used for the conditional variance process, and again the sum of the estimated coefficients is found to be close to unity.

Example 5.3: Conditional variance and the risk premium in the foreign exchange market

The evidence provided by Hansen and Hodrick (1980) for the rejection of the 'simple' efficiency hypothesis in foreign exchange markets, which was discussed in example 5.1, found that such rejection was often due to the intercept α_0 being non-zero. This finding could be regarded as evidence of a risk premium, the presence of which would allow the forward rate to be a biased predictor of the future spot rate without sacrificing the notion of market efficiency. Of course, for this to be plausible, we must have an empirically tractable theory of a risk premium; for without such a theory, there is no way of empirically distinguishing between an inefficient market and a, perhaps time-varying, risk premium.

Although several theoretical models have been proposed that generate a risk premium in the foreign exchange market, it has been found to be extremely difficult to translate them into testable econometric models and, consequently, their empirical performance provides only weak support for a time-varying risk premium. Domowitz and Hakkio (1985) therefore present a GARCH-M generalisation of the model used in example 5.1 to investigate the possible presence of a risk premium that depends on the conditional variance of the forecast errors. The model is arrived at as follows. From example 5.1, the efficiency hypothesis states that the forward rate at time t, $f_{t,1}$, is an unbiased predictor of the future spot rate, s_{t+1}, where, as before, logarithms are used, but where now we set the forecast period at $k=1$ for convenience. Thus

$$s_{t+1}-f_{t,1}=u_{t+1},$$

where u_{t+1} is the one-period forecast error, which should be zero mean white noise under the efficiency hypothesis. This can equivalently be written as

$$\nabla s_{t+1}=(f_{t,1}-s_t)+u_{t+1},$$

which is then regarded as a restricted case of the GARCH-M model of equations (5.6)–(5.8) with $y_t=\nabla s_t$ and $\mathbf{x}_t=(f_{t-1,1}-s_{t-1})$: the restrictions being $m=r=0$, so that no lagged y's and $\mathbf{x}$'s appear in the equation for y_t and that the forecast error is serially uncorrelated; $\alpha_0=\delta=0$, so that there is zero risk premium; and $\beta_0=1$, so that forecasts are unbiased. Maintaining $\beta_0=1$ and u_t to be white noise, then $\alpha_0\neq 0$ and $\delta=0$ implies a non-zero but constant risk premium, while $\alpha_0\neq 0$ and $\delta\neq 0$ implies a time-varying risk premium.

The risk premium is given by $\alpha_0+\delta h_t$ (assuming $\lambda=1$ for convenience) and thus any change in it is due solely to changes in the conditional variance h_t: it can, nevertheless, be positive or negative and can switch signs, depending on the values of α_0 and δ. For example, if $\alpha_0<0$ and $\delta>0$, then for small forecast errors the risk premium will be negative (long positions in foreign currency require an expected loss), while for large forecast errors the risk premium may turn positive (long positions in forward foreign currency require an expected profit).

The model was fitted, with h_t assumed to follow an ARCH(4) process, i.e., $p=4$, $q=0$, to non-overlapping monthly data from June 1973 to August 1982 for five exchange rates *vis-à-vis* the US dollar: the UK, France, Germany, Japan and Switzerland. The null hypothesis of no risk premium ($\alpha_0=0, \beta_0=1$, and $\delta=0$) could be rejected for the UK and Japan, but not for France, Germany or Switzerland, although for this last currency it is only because the standard error of β_0 is so large that the null cannot be rejected, for the point estimate of β_0 is -1.092!

5.2.2 Exponential GARCH-M models

As has been outlined, the family of GARCH models has proved to be a popular method of modelling volatility in financial markets. Nelson (1991), however, discusses some limitations that the GARCH models have in this context. One such limitation follows from the finding, obtained by many researchers, that changes in stock return volatility are negatively correlated with returns themselves, i.e., volatility tends to rise in response to 'bad news' (returns lower than expected) and to fall in response to 'good news' (returns higher than expected). GARCH models, however, assume that only the magnitude and not the sign of unanticipated returns determines volatility, i.e., it is the squared innovations, ϵ_t^2, that affect h_t in equation (5.8), so that h_t is invariant to the sign of the ϵ_ts. This suggests that a model in which h_t responds asymmetrically to positive and negative innovations might be useful for asset pricing applications (cf. the asymmetric time series model introduced in chapter 4.5.4).

A second important limitation of GARCH models results from the non-negativity constraints placed on the parameters of the GARCH process to ensure that h_t remains non-negative for all t. These constraints imply that increasing ϵ_t^2 in any period increases h_{t+m} for all $m \geq 1$, thus ruling out random oscillatory behaviour in the h_t process. Furthermore, these non-negativity constraints can create problems when estimating GARCH models and it may be difficult to determine whether shocks in GARCH models are persistent or not.

Nelson (1991) presents an alternative to the GARCH model that meets these objections and hence may be more suitable for modelling conditional variances in asset returns. To ensure that h_t remains non-negative, but without imposing inequality constraints on the model parameters, Nelson assumes that the (natural) logarithm of h_t is a function of lagged U_ts, where $U_t = \epsilon_t / h_t$ is the standardised process introduced in chapter 4.3.2, i.e., for some suitable function g

$$\ln(h_t) = \gamma_0 + g(U_{t-1}) + \sum_{i=1}^{p} \gamma_i g(U_{t-1-i}) + \sum_{i=1}^{q} \phi_i \ln(h_{t-i}) + \vartheta \xi_t .$$

Nelson chooses to make $g(U_t)$ a linear combination of U_t and $|U_t|$

$$g(U_t) = \varphi_1 U_t + \varphi_2 [|U_t| - E|U_t|]$$

and shows that such a specification allows an asymmetric response by h_t to changes in U_t, and hence ϵ_t, permits cycling in h_t, since no restrictions are placed on the γ_is, and enables the persistence of shocks to the variance to be easily evaluated.

Nelson (1991) fits a model of this type to daily US stock returns from July 1962 to December 1987, the actual form being

$$\begin{aligned} y_t &= \alpha_0 + \alpha_1 y_{t-1} + \delta h_t + \epsilon_t \\ \ln(h_t) &= \gamma_0 + \varphi_1 U_{t-1} + \varphi_2 [|U_{t-1}| - E|U_{t-1}|] \\ &\quad + \phi_1 \ln(h_{t-1}) + \phi_2 \ln(h_{t-2}) + \vartheta \xi_t . \end{aligned}$$

Here ξ_t is a variable measuring the influence of non-trading periods on the conditional variance and U_t is assumed to be generated as iid draws from a *Generalised Error* distribution. The estimated risk premium δ is negative but insignificant. The estimates of φ_1 and φ_2 are significantly negative and positive respectively, and this provides an asymmetric relationship between returns and changes in volatility. All the major episodes of high volatility are associated with market drops and the estimates of the autoregressive parameters ϕ_1 and ϕ_2 imply a largest root of almost unity, indicating substantial persistence of shocks to the conditional variance.

It is apparent from this discussion that GARCH-M models are proving extremely important in modelling volatility, which is a key variable playing a central role in many areas of finance. It is also clear that empirical applications of the model in finance will continue to increase, as the survey by Bollerslev, Chou and Kroner (1992) admirably demonstrates.

5.3 Misspecification testing

The regression techniques developed in section 5.1 are based on the assumption that the model (5.1) is correctly specified, i.e., that the assumptions underlying the model are valid. If they are not, then some of the techniques can be invalidated. It is thus important to be able to test these assumptions: such tests are known as *misspecification tests* and we begin their development by rewriting (5.1) as

$$\begin{aligned} y_t &= \alpha_0 + \beta_0 \mathbf{x}_t + \sum_{i=1}^{m} (\alpha_i y_{t-i} + \beta_i \mathbf{x}_{t-i}) + u_t \\ &= \alpha_0 + \beta_0 \mathbf{x}_t + \sum_{i=1}^{m} \beta_i^* \mathbf{z}_{t-i} + u_t, \end{aligned} \tag{5.9}$$

where $\beta_i^* = (\alpha_i, \beta_i)$, so that $\beta = (\beta_0^*, \beta_1^*, \ldots, \beta_m^*)$.

5.3.1 *Choosing the maximum lag, m*

The estimation theory developed in section 5.1 is based on the assumption that the maximum lag, m, is known. If this is so, then the assumption of mixing, which lets the errors u_t exhibit both serial correlation and heterogeneity, still allows the LS estimate $\hat{\beta}_T$ to be consistent and asymptotically normal, although the associated covariance matrix is $\mathbf{D}_T = (\mathbf{X}'\mathbf{X}/T)^{-1}\tilde{\mathbf{V}}_T(\mathbf{X}'\mathbf{X}/T)^{-1}$, where the expression for $\tilde{\mathbf{V}}_T$ is given by equation (5.3). If m is chosen to be larger than its optimum but unknown value m^*, $\hat{\beta}_T$ will still be consistent and asymptotically normal, but multicollinearity problems will often arise. This is because as m increases, the same observed data $\{\mathbf{z}_t\}_1^T$ are required to provide more and more information about an increasing number of unknown parameters.

If, on the other hand, m is chosen to be 'too small', then the omitted lagged $\mathbf{z}_t$s will form part of the error term. If we assume that for the correct lag length m^*, u_t is a martingale difference, then the error term in the misspecified model will no longer be non-systematic relative to $(\mathbf{y}_{t-1}^0, \mathbf{x}_t^0)$ and hence will not be a martingale difference. This has the implication that $\hat{\beta}_T$ and $\hat{\sigma}_T^2$ are no longer consistent or asymptotically normal and because of this it is important to be able to test for $m < m^*$. Given that the 'true' model is

$$y_t = \alpha_0 + \beta_0 \mathbf{x}_t + \sum_{i=1}^{m^*} (\alpha_i y_{t-i} + \beta_i \mathbf{x}_{t-i}) + u_t,$$

the error term in the misspecified model can be written as

$$u_t^* = u_t + \sum_{i=m+1}^{m^*} (\alpha_i y_{t-i} + \beta_i \mathbf{x}_{t-i}).$$

This implies that $m < m^*$ can be tested using the null hypothesis $H_0 : \beta_{m+1}^* = \ldots = \beta_{m^*}^* = 0$. The Wald statistic for testing this null against the alternative that at least one of the vectors β_i^*, $m+1 \leq i \leq m^*$, is non-zero is $(m^* - m)(k+1)$ times the standard F-statistic based on a comparison of the residual sums of squares from the regressions with the maximum lag length set at m and m^* respectively. The asymptotically equivalent LM statistic can be computed as $T \cdot R^2$ from the auxiliary regression of $\hat{u}_t^*$ on $\mathbf{x}_t$ and $\mathbf{z}_{t-1}, \ldots, \mathbf{z}_{t-m^*}$, where $\hat{u}_t^*$ are the residuals from the estimation of (5.9). Both the Wald and LM tests will be asymptotically χ_q^2, where $q = (m^* - m)(k+1)$.

The above analysis has assumed that for the correct lag length m^*, u_t is a martingale difference. One consequence of incorrectly setting m to be less than m^* is that the residuals from the regression of (5.9) will be serially correlated. An alternative LM test is $T \cdot R^2$ from the regression of $\hat{u}_t^*$ on $\mathbf{x}_t$, $\mathbf{z}_{t-1}, \ldots, \mathbf{z}_{t-m}$, and $\hat{u}_{t-1}^*, \ldots, u_{t-m+m^*}^*$, which will be asymptotically $\chi_{m^*-m}^2$. This is strictly a test of residual serial correlation, and only an indirect test of lag length specification, but it points to the difficulty of distinguishing whether residual serial correlation is a consequence of an incorrect (too small) setting of the lag length m, or whether m is correct but, nevertheless, the error term is serially correlated: as we have seen, in the former case $\hat{\beta}_T$ will be inconsistent, whereas in the latter it will be consistent and asymptotically normal. For detailed discussion of this important distinction, see Spanos (1986).

5.3.2 Testing for normality, linearity and homoskedasticity

Although the assumption that the errors in (5.9) are normally distributed is not a crucial one in the context of the asymptotic theory developed in section 5.1, it is often of interest to examine its validity, particularly as many financial time series are observed to be non-normal. A popular test proposed by Jarque and Bera (1980) measures departures from normality in terms of the third and fourth moments, i.e., the skewness and kurtosis, of the residuals $\hat{u}_t$ from estimation of (5.9). Letting μ_3 and μ_4 be the third and fourth (central) moments of u_t, and defining $m_3=(\mu_3/\sigma^3)$ and $m_4=(\mu_4/\sigma^4)$ to be the moment measures of skewness and kurtosis, respectively, estimators of these measures are given by

$$\hat{m}_i=\left[\left(\frac{1}{T}\Sigma\hat{u}_t^i\right)\Big/\left(\frac{1}{T}\Sigma\hat{u}_t^2\right)^{i/2}\right], \qquad i=3,4.$$

The asymptotic distributions of these estimators under the null hypothesis of normality are

$$T^{\frac{1}{2}}\hat{m}_3 \overset{a}{\sim} N(0,6)$$
$$T^{\frac{1}{2}}(\hat{m}_4-3) \overset{a}{\sim} N(0,24),$$

and, since they are also asymptotically independent, the squares of their standardised forms can be added to obtain

$$\left[\frac{T}{6}\hat{m}_3^2+\frac{T}{24}(\hat{m}_4-3)^2\right] \overset{a}{\sim} \chi_2^2,$$

so that large values of this statistic would flag significant departures from normality.

The model (5.9) assumes that the conditional mean $E(y_t\,|\,\mathbf{y}_{t-1}^0,\mathbf{x}_t^0)$ is linear in $\mathbf{X}_t$. To test this assumption we may consider the null hypothesis

$$H_0\colon \mu_{yt}=E(y_t\,|\,\mathbf{y}_{t-1}^0,\mathbf{x}_t^0)=\mathbf{X}_t\beta,$$

which needs to be tested against the non-linear alternative

$$H_1\colon \mu_{yt}=h(\mathbf{X}_t).$$

If $h(\cdot)$ is assumed to take the form

$$h(\mathbf{X}_t)=\mathbf{X}_t\Xi+c_2\mu_{yt}^2+c_3\mu_{yt}^3+\ldots+c_n\mu_{yt}^n,$$

then Ramsey's (1969) RESET test for linearity is based on testing $H_0{:}c_2=c_3=\ldots=c_n=0$ against $H_1{:}c_i\neq 0$, $i=2,\ldots,n$. Its LM version is based on the auxiliary regression of $\hat{u}_t$ on $\mathbf{x}_t$, $\mathbf{z}_{t-1},\ldots,\mathbf{z}_{t-m}$, and $\hat{\mu}_{yt}^2,\ldots,\hat{\mu}_{yt}^n$, where $\hat{\mu}_{yt}=\hat{y}_t=\mathbf{X}_t\hat{\beta}$, so that $T\cdot R^2$ is asymptotically distributed as χ_n^2. If non-linearities are encountered then non-linear regression techniques will be required: these are developed in White and Domowitz (1984) and analysed in detail in Gallant and White (1988).

To test for departures from homoskedasticity (assuming no serial correlation), we may consider constructing a test based on the difference

$$(\mathbf{X}'\Omega\mathbf{X})-\sigma^2(\mathbf{X}'\mathbf{X}),$$

where $\Omega=\text{diag}\,(\sigma_1^2,\sigma_2^2,\ldots,\sigma_T^2)$. This can be expressed in the form

$$\sum_{t=1}^{T}(E(u_t^2)-\sigma^2)\mathbf{X}_t\mathbf{X}_t',$$

and a test for heteroskedasticity could be based on the statistic

$$T^{-1}\sum_{t=1}^{T}(\hat{u}_t^2-\sigma_T^2)\mathbf{X}_t\mathbf{X}_t'.$$

Given that this is symmetric, we can express the $\frac{1}{2}K(K-1)$, where again $K=(m+1)(k+1)$, different elements in the form

$$T^{-1}\sum_{t=1}^{T}(\hat{u}_t^2-\sigma_T^2)\Psi_t, \tag{5.10}$$

where

$$\Psi_t=(\psi_{1t},\psi_{2t},\ldots,\psi_{Jt})', \qquad \psi_{lt}=x_{it}x_{jt},$$
$$i\geq j,\ i,j=2,\ldots,K, \quad l=1,2,\ldots,J, \quad J=\tfrac{1}{2}K(K-1),$$

the x_{it} being columns of $\mathbf{X}_t$. Although a test statistic can be based on (5.10), an asymptotically equivalent LM test (White, 1980) is the $T\cdot R^2$ statistic computed from the auxiliary regression of $\hat{u}_t^2$ on a constant and $\psi_{1t},\ldots,\psi_{Jt}$, which is asymptotically distributed as χ_J^2. Note, however, that the constant in the original regression (5.9) should not be involved in defining the ψ_{lt}s in the auxiliary regression, since the inclusion of such regressors would lead to perfect collinearity

This test, of course, does not propose any alternative form of heteroskedasticity. If such information is available, for example, that the errors follow an ARCH process, then tests specifically tailored to the alternative can be constructed: in the ARCH case the appropriate LM test is $T\cdot R^2$ from the regression of $\hat{u}_t^2$ on a constant and lags of $\hat{u}_t^2$ (cf. the testing of ARCH in chapter 4.4).

We note finally that throughout the analysis we have assumed that the parameter vector β is *time invariant*. Parameter time dependency may occur in many different forms and testing for departures from parameter time invariance is not easy. Spanos (1986, chapter 21, sections 5 and 6) provides an excellent discussion of this topic that is applicable to the class of models being considered here. Misspecification tests are now becoming available as standard options on econometric packages: both *MICROFIT 3.0* and *MICROTSP 7.0*, for example, allow a battery of tests to be carried out, including a range of parameter stability assessments.

Example 5.4: Testing the CAPM

The Capital Asset Pricing Model (CAPM) is an important asset pricing theory in financial economics and has been the subject of considerable econometric research. An excellent exposition of the derivation of the model which, as we have noted earlier in section 2.1, postulates a linear relationship between the expected risk and return of holding a portfolio of financial assets can be found in Berndt (1991, chapter 2), who also considers many of the econometric issues involved in the empirical implementation of the model.

The simple linear relationship between a small portfolio's return, r_p, and its associated risk, measured by the standard deviation of returns, σ_p, can be written as

$$r_p - r_f = (\sigma_p/\sigma_m)\cdot(r_m - r_f), \tag{5.11}$$

where r_m and σ_m are the returns on the overall market portfolio and the standard deviation of such returns, respectively, and r_f is the return on a risk-free asset. The term $r_p - r_f$ is thus the risk premium for portfolio p, while $r_m - r_f$ is the overall market's risk premium. Denoting these risk premia as y and x, respectively, letting $\beta = \sigma_p/\sigma_m$, and adding an intercept term α and a stochastic error term u, the latter reflecting the effects of specific (unsystematic) and diversible risk, the CAPM model becomes the simple linear regression

$$y = \alpha + \beta x + u. \tag{5.12}$$

The LS estimate of the slope coefficient β is $\hat{\beta} = Cov(x,y)/V(x)$, which is equivalent to σ_{pm}/σ_m^2, where σ_{pm} is the covariance between portfolio p and the market portfolio: this is known as the 'investment beta' for portfolio p and measures the sensitivity of the return on the portfolio to variation in the returns on the market portfolio. Portfolios having $\hat{\beta}$s in excess of unity are thus relatively risky, while those with $\hat{\beta}$s less than unity are much less sensitive to market movements.

LS estimation of the CAPM regression from observed time series $\{y_t, x_t\}_1^T$ is, of course, trivial: however, in this time series context the underlying CAPM theory requires certain assumptions to hold. Specifically, we must assume that the risk premia are stationary, normally distributed and serially uncorrelated, in which case the error process $\{u_t\}_1^T$ will be *NID*. Note also that the intercept α has been included without any justification: it does *not* appear in the original CAPM expression (5.11). The CAPM theory thus provides the testable hypothesis $\alpha = 0$, along with the following implications: the residuals of the regression (5.12) should be serially uncorrelated, homoskedastic and normal, the systematic relationship between y and x should be linear, and the estimate of β should be time invariant.

The empirical performance of the CAPM was investigated using the data set provided by Berndt (1991, chapter 2), which contains monthly returns from January 1978 to December 1987 on seventeen US companies plus a monthly value-weighted composite market return and a monthly risk-free return. Treating each companies' risk premia, calculated as the difference between the company return and the risk-free return, as a separate portfolio enabled seventeen CAPM regressions of the form (5.12) to be estimated and these are reported in table 5.1.

Only three of the estimated regressions survive the battery of misspecification tests unscathed: those for CONED, DELTA and MOTOR (see Berndt for the actual companies associated with these variable names). Little evidence of serial correlation or heteroskedasticity is found in the residuals, the absence of the latter, particularly in ARCH form, being somewhat surprising given previous findings in this area (see, for example, Morgan and Morgan, 1987). Much greater evidence of non-linearity, non-normality and parameter non-constancy is encountered. This is probably a reflection of the now commonly held view that 'betas' are time-varying and are better modelled within a GARCH-M framework.

Although both point estimates and standard errors of α and β may well be affected by these apparent misspecifications, it seems unlikely that the overall assessment of these values will alter much: estimates of α are invariably very small and insignificantly different from zero, while estimates of β are usually between zero and unity and certainly not significantly greater than unity for any stock. The R^2 statistic measures the market (systematic) portion of total risk and ranges from almost zero to just over 0.40, with typical values being in the region of 0.30.

5.4 The multivariate linear regression model

An immediate extension of the regression model (5.1) is to replace the 'dependent' variable y_t by a vector, say $\mathbf{y}_t=(y_{1t},\dots,y_{nt})'$, so that we now have the *multivariate (dynamic) regression model*

$$y_t=\mathbf{C}+\sum_{i=1}^{m}\mathbf{A}_i'\mathbf{y}_{t-i}+\sum_{i=0}^{m}\mathbf{B}_i'\mathbf{x}_{t-i}+\mathbf{u}_t,\qquad m+1\leq t\leq T, \tag{5.13}$$

where $\mathbf{C}$ is an $n\times 1$ vector of constants, $\mathbf{A}_1,\dots,\mathbf{A}_m$ are $n\times n$ matrices of lag coefficients, $\mathbf{B}_0,\mathbf{B}_1,\dots,\mathbf{B}_m$ are $k\times n$ coefficient matrices, and $\mathbf{u}_t$ is an $n\times 1$ vector of errors having the properties

$$E(\mathbf{u}_t)=E(\mathbf{u}_t\,|\,\mathbf{Y}_{t-1}^0,\mathbf{x}_t^0)=0$$

and

$$E(\mathbf{u}_t\mathbf{u}_s')=E\{E(\mathbf{u}_t\mathbf{u}_s'\,|\,\mathbf{Y}_{t-1}^0,\mathbf{x}_t^0)\}=\begin{cases}\Omega & t=s\\ \mathbf{0} & t\neq s\end{cases},$$

where

$$\mathbf{Y}_{t-1}^0=(\mathbf{y}_{t-1},\mathbf{y}_{t-2},\dots,\mathbf{y}_1).$$

Table 5.1 *Estimates of the CAPM regression (5.12)*

Company	$\hat{\alpha}$	$\hat{\beta}$	R^2	*dw*	*NONLIN*	*NORM*	*HET*	*ARCH*	*CHOW*
BOISE	0.0031 (0.0068)	0.94 (0.10)	0.43	2.17	2.91	5.03	1.66	8.69*	2.35
CITCRP	0.0025 (0.0062)	0.67 (0.09)	0.32	1.84	0.40	1.34	8.09*	2.50	6.69*
CONED	0.0110 (0.0046)	0.09 (0.07)	0.02	2.15	0.74	1.22	0.19	5.04	0.02
CONTIL	−0.0132 (0.0131)	0.73 (0.19)	0.11	2.07	0.44	2287*	0.10	0.06	2.62
DATGEN	−0.0067 (0.0098)	1.03 (0.14)	0.31	2.08	6.99*	5.32	2.69	0.15	2.40
DEC	0.0068 (0.0074)	0.85 (0.11)	0.34	2.14	0.71	9.72*	0.65	16.03*	5.26
DELTA	0.0014 (0.0083)	0.49 (0.12)	0.12	1.99	0.01	2.59	0.09	2.46	4.08
GENMIL	0.0078 (0.0058)	0.27 (0.08)	0.08	2.08	0.14	2.60	0.94	2.16	15.31*
GERBER	0.0051 (0.0071)	0.63 (0.10)	0.24	2.25	8.09*	7.20*	0.17	1.72	6.58*
IBM	−0.0005 (0.0046)	0.46 (0.07)	0.28	1.88	0.06	1.12	0.02	3.06	6.72*
MOBIL	0.0042 (0.0059)	0.72 (0.09)	0.37	2.09	0.55	35.6*	0.21	1.68	0.49

MOTOR	0.0069 (0.0083)	0.10 (0.12)	0.01	1.86	0.90	2.17	0.77	0.73	1.97
PANAM	−0.0086 (0.0112)	0.74 (0.16)	0.15	2.21	0.51	11.4*	0.25	4.46	0.14
PSNH	−0.0126 (0.0100)	0.21 (0.15)	0.02	1.88	0.25	95.0*	0.66	10.46*	0.25
TANDY	0.0107 (0.0097)	1.05 (0.14)	0.32	1.89	3.23	6.32*	0.18	0.66	0.13
TEXACO	0.0007 (0.0062)	0.61 (0.09)	0.28	2.02	0.00	13.1*	0.04	3.10	0.14
WEYER	−0.0031 (0.0059)	0.82 (0.09)	0.43	2.29*	1.74	1.37	0.65	9.97*	9.82*
Asymptotic distribution					χ^2_1	χ^2_2	χ^2_1	χ^2_3	χ^2_2
Critical 0.05 Value				1.72 2.28	3.84	5.99	3.84	7.81	5.99

Notes:
*: Significant at 0.05 level.
(...): Conventional standard error.
dw: Durbin–Watson statistic.
NONLIN: Ramsey's RESET test for functional form, calculated from the regression of $\hat{u}_t$ on x_t and $\hat{y}_t^2$.
NORM: Barque and Jera test for normality.
HET: Test for heteroskedasticity, calculated from the regression of $\hat{u}_t^2$ on a constant and $\hat{y}_t^2$.
CHOW: Chow's (1960) test for coefficient stability; break point taken to be December 1984.

In matrix form, we have

$$\mathbf{Y}=\mathbf{X}^*\mathbf{B}+\mathbf{U},$$

where

$$\mathbf{Y}=(\mathbf{y}_{m+1},\ldots,\mathbf{y}_T)'$$
$$\mathbf{X}^*=(\mathbf{X}^*_{m+1},\ldots,\mathbf{X}^*_T)'$$
$$\mathbf{X}^*_t=(1,\mathbf{y}_{t-1},\ldots,\mathbf{y}_{t-m},\mathbf{x}_t,\ldots,\mathbf{x}_{t-m})$$
$$\mathbf{U}=(\mathbf{u}_{m+1},\ldots,\mathbf{u}_T)'$$

and

$$\mathbf{B}=(\mathbf{C}',\mathbf{A}'_1,\ldots,\mathbf{A}'_m,\mathbf{B}'_0,\ldots,\mathbf{B}'_m).$$

The estimation theory for this model is basically a multivariate extension of that developed for the univariate case ($n=1$) above. For example, the LS and (approximate) ML estimator of **B** is

$$\hat{\mathbf{B}}=(\mathbf{X}^{*\prime}\mathbf{X}^*)^{-1}\mathbf{X}^{*\prime}\mathbf{Y},$$

while the ML estimator of Ω is

$$\hat{\Omega}=T^{-1}\hat{\mathbf{U}}'\hat{\mathbf{U}},\quad \hat{\mathbf{U}}=\mathbf{Y}-\mathbf{X}^{*\prime}\hat{\mathbf{B}}.$$

Spanos (1986, chapter 24) considers this model in some detail, presenting misspecification tests that are essentially multivariate extensions of those outlined in section 5.3.

Example 5.5: Multivariate tests of the CAPM

Since the publication of Gibbons (1982), multivariate tests of the CAPM have been the subject of considerable research. The multivariate CAPM can be analysed empirically within the framework of the multivariate regression model: by letting $\mathbf{y}_t$ be the vector of n excess asset returns at time t and x_t be the excess market return at time t, the model can be written as

$$\mathbf{y}_t=\mathbf{C}+\mathbf{B}x_t+\mathbf{u}_t,$$

where **C** and **B** are $n\times 1$ vectors of parameters and the error $\mathbf{u}_t$ has the properties of the error in equation (5.13). The CAPM model thus imposes the n restrictions that the intercepts in each asset return equation are zero, i.e., $\mathbf{C}=\mathbf{0}$. MacKinlay (1987, see also Gibbons, Ross and Shanken, 1989) shows that this hypothesis can be tested using the statistic

$$\theta=\frac{(T-n-1)T}{(T-2)n}\left[1+\frac{\bar{x}^2}{s_x^2}\right]^{-1}\mathbf{C}'\hat{\Omega}^{-1}\mathbf{C},$$

where $\bar{x}$ and s_x^2 are the sample mean and variance of x_t. Under H_0:$\mathbf{C}=0$, θ is distributed as F with n and $T-n-1$ degrees of freedom.

The $n=17$ assets considered separately in example 5.4 were reexamined in this multivariate framework. Of course, since the same (single) regressor appears in each equation, slope and intercept estimates are the same as the single equation LS estimates. A test of $\mathbf{C}=\mathbf{0}$ produces a θ value of 0.71, with an associated marginal significance level of 0.79: not surprisingly, given the intercept estimates presented in table 5.1, we cannot reject the null that all intercepts are zero, in accordance with the predictions of the CAPM, although we should emphasise that none of the misspecifications uncovered in the individual asset models in example 5.4 have been tackled here.

5.5 Vector autoregressions

5.5.1 Concepts of exogeneity and causality

Throughout the development of the various forms of regression models encountered so far in this chapter we have made the assumption that $\mathbf{y}_t$ is a function of past values of itself and present and past values of $\mathbf{x}_t$. More precisely, we have been assuming that $\mathbf{x}_t$ is *weakly exogenous*: the stochastic structure of $\mathbf{x}_t$ contains no information that is relevant for the estimation of the parameters of interest, $\mathbf{B}$ and Ω. Formally, $\mathbf{x}_t$ will be weakly exogenous if, when the joint distribution of $\mathbf{z}_t=(\mathbf{y}_t,\mathbf{x}_t)$, conditional on the past, is factorised as the conditional distribution of $\mathbf{y}_t$ given $\mathbf{x}_t$ times the marginal distribution of $\mathbf{x}_t$; (a) the parameters of these conditional and marginal distributions are not subject to cross-restrictions, and (b) the parameters of interest can be uniquely determined from the parameters of the conditional model alone. Under these conditions $\mathbf{x}_t$ may be treated 'as if' it were determined outside the conditional model for $\mathbf{y}_t$. For more details on weak exogeneity, see Engle, Hendry and Richard (1983) and Spanos (1986, chapter 19). Because it is a condition on parameters, rather than a restriction on joint probability distributions, it is usual to treat weak exogeneity as a non-directly testable assumption, although Spanos, for example, considers possible ways in which the assumption can be tested indirectly.

While the weak exogeneity of $\mathbf{x}_t$ allows efficient estimation of $\mathbf{B}$ and Ω without any reference to the stochastic structure of $\mathbf{x}_t$, the marginal distribution of $\mathbf{x}_t$, while not containing $\mathbf{y}_t$, will contain $\mathbf{Y}^0_{t-1}$, and the possible presence of lagged $\mathbf{y}_t$s can lead to problems when attempting to predict $\mathbf{y}_t$. In order to be able to treat the $\mathbf{x}_t$s as given when predicting $\mathbf{y}_t$, we need to ensure that no *feedback* exists from $\mathbf{Y}^0_{t-1}$ to $\mathbf{x}_t$: the absence of such feedback is equivalent to the statement that $\mathbf{y}_t$ *does not Granger-cause* $\mathbf{x}_t$. Weak exogeneity supplemented with Granger non-causality is called *strong exogeneity*.

Unlike weak exogeneity, Granger non-causality is directly testable (the original reference to this concept of causality is Granger, 1969). To investigate such tests, and to relate Granger non-causality to yet another concept of exogeneity, we need to introduce the *dynamic structural equation model* (DSEM) and the *vector autoregressive* (VAR) process. The DSEM extends the multivariate regression model in two directions: first, by allowing 'simultaneity' between the 'endogenous' variables in $\mathbf{y}_t$ and, second, explicitly considering the process generating the 'exogenous' variables $\mathbf{x}_t$. We thus have

$$\mathbf{A}_0\mathbf{y}_t=\sum_{i=1}^{m}\mathbf{A}_i'\mathbf{y}_{t-i}+\sum_{i=0}^{m}\mathbf{B}_i'\mathbf{x}_{t-i}+\mathbf{u}_{1t} \tag{5.14}$$

and

$$\mathbf{x}_t=\sum_{i=1}^{m}\mathbf{C}_i'\mathbf{x}_{t-i}+\mathbf{u}_{2t}. \tag{5.15}$$

The simultaneity of the model is a consequence of $\mathbf{A}_0 \neq \mathbf{I}_n$. The errors $\mathbf{u}_{1t}$ and $\mathbf{u}_{2t}$ are assumed to be jointly independent processes, which could be serially correlated but will be assumed here to be white noise, and intercept vectors are omitted for simplicity: see Mills (1990, chapter 14) and, particularly, Lutkepohl (1991) for a more general development. The identification conditions for the set of *structural* equations (5.14) are summarised in Hendry, Pagan and Sargan (1984), while (5.15) shows that $\mathbf{x}_t$ is generated by an mth order VAR process, in which current values of $\mathbf{x}$ are functions of m past values of $\mathbf{x}$ *only*.

If, in the DSEM (5.14), $E(\mathbf{u}_{1t}\mathbf{x}_{t-s})=0$ for *all* s, $\mathbf{x}_t$ is said to be *strictly* exogenous. Strict exogeneity is useful because no information is lost by limiting attention to distributions conditional on $\mathbf{x}_t$, which will usually result in considerable simplifications in statistical inference: for example, IV techniques may be used in the presence of serially correlated disturbances. A related concept is that of a variable being *predetermined*: a variable is predetermined if all its current and past values are independent of the current error $\mathbf{u}_{1t}$. If $\mathbf{x}_t$ is strictly exogenous, then it will also be predetermined, while if $E(\mathbf{u}_{1t}\mathbf{y}_{t-s})=0$, for $s>0$, then $\mathbf{y}_{t-s}$ will be predetermined as well.

In many cases, strictly exogenous variables will also be weakly exogenous in DSEMs, although one important class of exceptions is provided by rational expectations variables, in which behavioural parameters are generally linked to the distributions of exogenous variables. Similarly, predetermined variables will usually be weakly exogenous, except again in the case where there are cross-restrictions between behavioural parameters and the parameters of the distribution of the predetermined variables.

Strict exogeneity can be tested in DSEMs by using the *final form*, in which each endogenous variable is expressed as an infinite distributed lag of the exogenous variables

$$\mathbf{y}_t = \sum_{i=0}^{\infty} \mathbf{J}_i \mathbf{x}_{t-i} + \mathbf{e}_t,$$

where the $\mathbf{J}_i$ matrices are functions of the $\mathbf{A}_i$s and $\mathbf{B}_i$s and where $\mathbf{e}_t$ is a stochastic process possessing a VAR representation and having the property that $E(\mathbf{e}_t \mathbf{x}_{t-s}) = 0$ for all s. Geweke (1978) proves that, in the regression of $\mathbf{y}_t$ on *all* current, lagged, and future values of $\mathbf{x}_t$,

$$\mathbf{y}_t = \sum_{i=-\infty}^{\infty} \mathbf{K}_i \mathbf{x}_{t-i} + \mathbf{e}_t, \tag{5.16}$$

then if, and only if, the coefficients on *future* values of $\mathbf{x}_t$, i.e., $\mathbf{x}_{t-s}$, $s < 0$, are equal to zero will there exist a DSEM relating $\mathbf{x}_t$ and $\mathbf{y}_t$ in which $\mathbf{x}_t$ is strictly exogenous. An equivalent test is based on the regression

$$\mathbf{x}_t = \sum_{i=1}^{\infty} \mathbf{E}_{2i} \mathbf{x}_{t-i} + \sum_{i=1}^{\infty} \mathbf{F}_{2i} \mathbf{y}_{t-i} + \mathbf{w}_t, \tag{5.17}$$

in which $E(\mathbf{y}_{t-i} \mathbf{w}_t') = 0$ for all t and $s > 0$. Geweke proves that $\mathbf{x}_t$ will be strictly exogenous in a DSEM relating $\mathbf{x}_t$ and $\mathbf{y}_t$ if, and only if, the coefficient matrices $\mathbf{F}_{2i}$, $i = 1,2,\ldots,$ are all zero.

Strict exogeneity is intimately related to Granger non-causality. Indeed, the two tests for strict exogeneity of $\mathbf{x}_t$ above can also be regarded as tests for $\mathbf{y}_t$ not Granger-causing $\mathbf{x}_t$. The two concepts are *not* equivalent, however. As Geweke (1984) points out, if $\mathbf{x}_t$ is strictly exogenous in the DSEM (5.14), then $\mathbf{y}_t$ does not Granger-cause $\mathbf{x}_t$, where $\mathbf{y}_t$ is endogenous in that model. However, if $\mathbf{y}_t$ does not Granger-cause $\mathbf{x}_t$, then there exists *a* DSEM with $\mathbf{y}_t$ endogenous and $\mathbf{x}_t$ strictly exogenous, in the sense that there will exist systems of equations formally similar to (5.14), *but* none of these systems need necessarily satisfy the overidentifying restrictions of the specific model. This implies that tests for the absence of a causal ordering can be used to refute the strict exogeneity specification in a given DSEM, but such tests cannot be used to establish it.

Furthermore, as we have already discussed, statistical inference may be carried out conditionally on a subset of variables that are not strictly exogenous: all that we require is that they be weakly exogenous. Thus, unidirectional Granger causality is neither necessary nor sufficient for inference to proceed conditional on a subset of variables.

5.5.2 Tests of Granger causality and measures of feedback

To develop operational tests of Granger causality and strict exogeneity, we now consider the $g=n+k$ dimensional vector $\mathbf{z}_t=(\mathbf{y}_t,\mathbf{x}_t)$, which we assume has the following mth order VAR representation

$$\mathbf{z}_t=\sum_{i=1}^{m}\pi_i\mathbf{z}_{t-i}+\mathbf{v}_t, \tag{5.18}$$

where

$$E(\mathbf{v}_t)=E(\mathbf{v}_t\,|\,\mathbf{Z}_{t-1}^0)=0,$$

$$E(\mathbf{v}_t\mathbf{v}_s')=E\{E(\mathbf{v}_t\mathbf{v}_s'\,|\,\mathbf{Z}_{t-1}^0)\}=\begin{cases}\Sigma_{\mathbf{v}}, & t=s\\ \mathbf{0}, & t\neq s\end{cases}$$

and

$$\mathbf{Z}_{t-1}^0=(\mathbf{z}_{t-1},\mathbf{z}_{t-2},\ldots,\mathbf{z}_1).$$

The VAR of equation (5.18) can be partitioned as

$$\mathbf{y}_t=\sum_{i=1}^{m}\mathbf{C}_{2i}\mathbf{x}_{t-i}+\sum_{i=1}^{m}\mathbf{D}_{2i}\mathbf{y}_{t-i}+\mathbf{v}_{1t} \tag{5.19}$$

$$\mathbf{x}_t=\sum_{i=1}^{m}\mathbf{E}_{2i}\mathbf{x}_{t-i}+\sum_{i=1}^{m}\mathbf{F}_{2i}\mathbf{y}_{t-i}+\mathbf{v}_{2t}, \tag{5.20}$$

where $\mathbf{v}_t'=(\mathbf{v}_{1t}',\mathbf{v}_{2t}')$ and where $\Sigma_{\mathbf{v}}$ is correspondingly partitioned as

$$\Sigma_{\mathbf{v}}=\begin{pmatrix}\Sigma_{11} & \Sigma_{12}\\ \Sigma_{12}' & \Sigma_{22}\end{pmatrix}.$$

Here $\Sigma_{ij}=E(\mathbf{v}_{it}\mathbf{v}_{jt}')$, $i,j=1,2$, so that, although the error vectors $\mathbf{v}_{1t}$ and $\mathbf{v}_{2t}$ are each serially uncorrelated, they can be correlated with each other contemporaneously, although at no other lag. Given equations (5.19) and (5.20), $\mathbf{y}$ *does not Granger-cause* $\mathbf{x}$ if, and only if, $\mathbf{F}_{2i}\equiv 0$, for all i. An equivalent statement of this proposition is that $|\Sigma_{22}|=|\Sigma_2|$, where $\Sigma_2=E(\mathbf{w}_{2t}\mathbf{w}_{2t}')$, obtained from the 'restricted' regression

$$\mathbf{x}_t=\sum_{i=1}^{m}\mathbf{E}_{1i}\mathbf{x}_{t-i}+\mathbf{w}_{2t}. \tag{5.21}$$

Similarly, $\mathbf{x}$ *does not Granger-cause* $\mathbf{y}$ if, and only if, $\mathbf{C}_{2i}\equiv 0$, for all i or, equivalently, that $|\Sigma_{11}|=|\Sigma_1|$, where $\Sigma_1=E(\mathbf{w}_{1t}\mathbf{w}_{1t}')$, obtained from the regression

$$\mathbf{y}_t=\sum_{i=1}^{m}\mathbf{C}_{1i}\mathbf{y}_{t-i}+\mathbf{w}_{1t}. \tag{5.22}$$

If the system (5.19)–(5.20) is premultiplied by the matrix

$$\begin{bmatrix} \mathbf{I}_n & -\Sigma_{12}\Sigma_{22}^{-1} \\ -\Sigma'_{12}\Sigma_{11}^{-1} & \mathbf{I}_k \end{bmatrix},$$

then the first n equations of the new system can be written as

$$\mathbf{y}_t = \sum_{i=0}^{m} \mathbf{C}_{3i}\mathbf{x}_{t-i} + \sum_{i=1}^{m} \mathbf{D}_{3i}\mathbf{y}_{t-i} + \omega_{1t}, \tag{5.23}$$

where the error $\omega_{1t} = \mathbf{v}_{1t} - \Sigma_{12}\Sigma_{22}^{-1}\mathbf{v}_{2t}$, since it is uncorrelated with $\mathbf{v}_{2t}$, is also uncorrelated with $\mathbf{x}_t$. Similarly, the last k equations can be written as

$$\mathbf{x}_t = \sum_{i=1}^{m} \mathbf{E}_{3i}\mathbf{x}_{t-i} + \sum_{i=0}^{m} \mathbf{F}_{3i}\mathbf{y}_{t-i} + \omega_{2t}. \tag{5.24}$$

Denoting $\Sigma_{\omega i} = E(\omega_{it}\omega'_{it})$, $i = 1,2$, there is *instantaneous causality* between $\mathbf{y}$ and $\mathbf{x}$ if, and only if, $\mathbf{C}_{30} \neq 0$ and $\mathbf{E}_{30} \neq 0$ or, equivalently, $|\Sigma_{11}| > |\Sigma_{\omega 1}|$ and $|\Sigma_{22}| > |\Sigma_{\omega 2}|$.

Given this framework, Geweke (1982) defines a *measure of linear feedback from* $\mathbf{y}$ *to* $\mathbf{x}$ as

$$F_{\mathbf{y}\to\mathbf{x}} = \ln(|\Sigma_2| / |\Sigma_{22}|),$$

so that the statement '$\mathbf{y}$ does not cause $\mathbf{x}$' is equivalent to $F_{\mathbf{y}\to\mathbf{x}} = 0$. Symmetrically, $\mathbf{x}$ does not cause $\mathbf{y}$ if, and only if, the *measure of linear feedback from* $\mathbf{x}$ *to* $\mathbf{y}$,

$$F_{\mathbf{x}\to\mathbf{y}} = \ln(|\Sigma_1| / |\Sigma_{11}|),$$

is zero. The existence of instantaneous causality between $\mathbf{y}$ and $\mathbf{x}$ amounts to a non-zero measure of *instantaneous linear feedback*

$$F_{\mathbf{x}\cdot\mathbf{y}} = \ln(|\Sigma_{11}| / |\Sigma_{\omega 1}|) = \ln(|\Sigma_{22}| / |\Sigma_{\omega 2}|).$$

A concept closely related to the idea of linear feedback is that of *linear dependence*, a measure of which is given by

$$F_{\mathbf{x},\mathbf{y}} = \ln(|\Sigma_1| / |\Sigma_{\omega 1}|) = \ln(|\Sigma_2| / |\Sigma_{\omega 2}|).$$

From these measures it is easily seen that

$$F_{\mathbf{x},\mathbf{y}} = F_{\mathbf{y}\to\mathbf{x}} + F_{\mathbf{x}\to\mathbf{y}} + F_{\mathbf{x}\cdot\mathbf{y}},$$

so that linear dependence can be decomposed additively into the three forms of feedback. Absence of a particular causal ordering is then equivalent to one of these feedback measures being zero.

To obtain estimates of these measures, we shall suppose that each of the regressions (5.19)–(5.24) have been estimated by LS (this will yield consistent and efficient estimates) and the following matrices formed

$$\hat{\Sigma}_i = (T-m)^{-1} \sum_{t=m+1}^{T} \hat{\mathbf{w}}_{it}\hat{\mathbf{w}}'_{it},$$

$$\hat{\Sigma}_{ii} = (T-m)^{-1} \sum_{t=m+1}^{T} \hat{\mathbf{v}}_{it}\hat{\mathbf{v}}'_{it},$$

$$\hat{\Sigma}_{\omega i} = (T-m)^{-1} \sum_{t=m+1}^{T} \hat{\omega}_{it}\hat{\omega}'_{it},$$

for $i=1,2$, where $\hat{\mathbf{w}}_{it}$ is the vector of LS residuals corresponding to the error vector $\mathbf{w}_{it}$, etc. From these estimates we can then compute the various feedback measures: for example

$$\hat{F}_{\mathbf{y}\to\mathbf{x}} = \ln(|\hat{\Sigma}_2|/|\hat{\Sigma}_{22}|).$$

It then follows that the LR test statistic of the null hypothesis H_{01}: $F_{\mathbf{y}\to\mathbf{x}}=0$ (**y** does not Granger-cause **x**) is

$$\mathscr{L}\mathscr{R}: (T-m)\hat{F}_{\mathbf{y}\to\mathbf{x}} \sim \chi^2_{nkm}.$$

Similarly, the null H_{02}: $F_{\mathbf{x}\to\mathbf{y}}=0$ is tested by

$$(T-m)\hat{F}_{\mathbf{x}\to\mathbf{y}} \sim \chi^2_{nkm},$$

and H_{03}: $F_{\mathbf{x}\cdot\mathbf{y}}=0$ by

$$(T-m)\hat{F}_{\mathbf{x}\cdot\mathbf{y}} \sim \chi^2_{nk}.$$

Since these are tests of nested hypotheses, $\hat{F}_{\mathbf{y}\to\mathbf{x}}$, $\hat{F}_{\mathbf{x}\to\mathbf{y}}$ and $\hat{F}_{\mathbf{x}\cdot\mathbf{y}}$ are asymptotically independent. All three restrictions can be tested at once since

$$(T-m)\hat{F}_{\mathbf{y},\mathbf{x}} \sim \chi^2_{nk(2m+1)}$$

on H_{04}: $F_{\mathbf{y},\mathbf{x}}=0$.

The corresponding Wald and LM statistics testing, for example, H_{01}: $F_{\mathbf{y}\to\mathbf{x}}=0$, are

$$\mathscr{W}: (T-m)[\mathrm{tr}(\hat{\Sigma}_2\hat{\Sigma}_{22}^{-1})-n] \sim \chi^2_{nkm}$$

and

$$\mathscr{L}\mathscr{M}: (T-m)[n-\mathrm{tr}(\hat{\Sigma}_{22}\hat{\Sigma}_2^{-1})] \sim \chi^2_{nkm},$$

respectively. Geweke (1982) also considers the construction of confidence intervals. An approximate 95 per cent confidence interval for $F_{\mathbf{y}\to\mathbf{x}}$, for example, is given by

$$\left\{\left[\left(\hat{F}_{\mathbf{y}\to\mathbf{x}} - \frac{nkm-1}{3(T-m)}\right)^{\frac{1}{2}} - \frac{1.96}{\sqrt{T-m}}\right]^2 - \frac{2nkm+1}{3(T-m)},\right.$$

$$\left[\left(\hat{F}_{\mathbf{y}\to\mathbf{x}} - \frac{nkm-1}{3(T-m)}\right)^{\frac{1}{2}} + \frac{1.96}{\sqrt{T-m}}\right]^2 - \frac{2nkm+1}{3(T-m)}\Bigg\}.$$

Other tests of causality can be constructed, but they tend to require considerably more computation and, in any event, simulation studies carried out by a variety of authors reach a consensus that inference should be carried out using the procedures detailed above, these being found to combine the greatest reliability with computational ease.

5.5.3 Determining the order of a VAR

The measures of feedback and tests of causality developed above assume that the order m of the underlying VAR is known. In practice, of course, m will be unknown and must be determined empirically. A traditional tool for determining the order is to use a sequential testing procedure. If we have the g-dimensional VAR

$$\mathbf{z}_t = \sum_{i=1}^{m} \pi_i \mathbf{z}_{t-i} + \mathbf{v}_t,$$

from which the ML estimate of $\Sigma_{\mathbf{v}}$ is

$$\hat{\Sigma}_{\mathbf{v},m} = T^{-1}\mathbf{V}_m\mathbf{V}'_m,$$

where $\mathbf{V}_m = [\hat{\mathbf{v}}_{m+1}, \ldots, \hat{\mathbf{v}}_T]$ is the matrix of residuals obtained by LS estimation of the mth order VAR [VAR(m)], then the LR statistic for testing m against l, $l < m$, is

$$\mathscr{LR}:\ T\ln(|\hat{\Sigma}_{\mathbf{v},l}|/|\hat{\Sigma}_{\mathbf{v},m}|) \sim \chi^2_{g(m-l)}.$$

The corresponding W and LM statistics are

$$\mathscr{W}:\ T[\mathrm{tr}(\hat{\Sigma}_{\mathbf{v},l}\hat{\Sigma}^{-1}_{\mathbf{v},m}) - g] \sim \chi^2_{g(m-l)},$$

and

$$\mathscr{LM}:\ T[g - \mathrm{tr}(\hat{\Sigma}_{\mathbf{v},m}\hat{\Sigma}^{-1}_{\mathbf{v},l})] \sim \chi^2_{g(m-l)}.$$

Other procedures are based upon minimising some objective function and are essentially multivariate analogues of those discussed in example 2.3. Lutkepohl (1985) provides a comparison of the alternatives and reports various simulation experiments designed to investigate their empirical performance. The objective function that appears to be the most favoured is Schwarz' (1978) multivariate *BIC* Criterion, which is defined here as

$$\mathscr{SC}(j) = \ln|\hat{\Sigma}_{\mathbf{v},j}| + g^2 jT^{-1}\ln T \qquad j = 0, 1, \ldots, m,$$

where m is now the maximum order considered. This can be shown to provide a consistent estimate of the correct lag order and Lutkepohl finds that it also chooses the correct order most often, and the resulting VAR models provide the best forecasts, in a Monte Carlo comparison of objective functions.

After a tentative model has been specified using one of these procedures, checks on its adequacy may be carried out. These are analogous to the diagnostic checks used for univariate models and might involve overfitting and testing the significance of the extra parameters, plotting standardised residuals against time and analysing the estimated cross-correlation matrices of the residual series. Multivariate portmanteau and LM statistics are also available, but with vector time series there is probably no substitute for detailed inspection of the residual correlation structure for revealing subtle relationships which may indicate important directions of model improvement.

5.6 Variance decompositions and innovation accounting

A concise representation of the VAR(m) model is obtained by using lag operator notation

$$\pi(B)\mathbf{z}_t = \mathbf{v}_t,$$

where

$$\pi(B) = \mathbf{I} - \pi_1 B - \pi_2 B^2 - \ldots - \pi_m B^m.$$

Analogous to the univariate case, the vector MA representation of $\mathbf{z}_t$ is

$$\mathbf{z}_t = \pi^{-1}(B)\mathbf{v}_t = \psi(B)\mathbf{v}_t = \mathbf{v}_t + \sum_{i=1}^{\infty} \psi_i \mathbf{v}_{t-i}. \tag{5.25}$$

In this set up, no distinction is made between endogenous and (strictly) exogenous variables, so the ψ_i matrices can be interpreted as the *dynamic multipliers* of the system since they represent the model's response to a unit shock in each of the variables. The response of z_i to a unit shock in z_j is therefore given by the sequence, known as the *impulse response function*,

$$\psi_{ij,1}, \psi_{ij,2}, \psi_{ij,3}, \ldots,$$

where $\psi_{ij,k}$ is the ijth element of the matrix ψ_k. If a variable, or block of variables, are strictly exogenous, then the implied zero restrictions ensure that these variables do not react to a shock to any of the endogenous variables. Recall, however, that $E(\mathbf{v}_t\mathbf{v}_t') = \Sigma_\mathbf{v}$, so that the components of $\mathbf{v}_t$ are contemporaneously correlated. If these correlations are high, simulation of a shock to z_j, while all other components of **z** are held constant, could be misleading.

However, if we define the lower triangular matrix **S** such that $\mathbf{SS}'=\Sigma_v$ and $\mathbf{n}_t=\mathbf{S}^{-1}\mathbf{v}_t$, then $E(\mathbf{n}_t\mathbf{n}_t')=\mathbf{I}_g$ and we can renormalise the MA representation (5.25) into the *recursive* form

$$\mathbf{z}_t=\sum_{i=0}^{\infty}\psi_i^*\mathbf{n}_{t-i},$$

where $\psi_i^*=\psi_i\mathbf{S}^{-1}$ (so that $\psi_0^*=\mathbf{S}^{-1}$ is lower triangular). The impulse response function of z_i to a unit shock in z_j is then given by the sequence

$$\psi_{ij,0}^*,\ \psi_{ij,1}^*,\ \psi_{ij,2}^*,\ \ldots$$

The uncorrelatedness of the $\mathbf{n}_t$s allows the error variance of the $H+1$ step-ahead forecast of z_i to be decomposed into components accounted for by these shocks, or innovations; hence the phrase coined by Sims (1981) for this technique, that of *innovation accounting*. In particular, the components of this error variance accounted for by innovations to z_j is given by

$$\sum_{h=0}^{H+1}\psi_{ij,h}^{*2}.$$

For large H, this *variance decomposition* allows the isolation of those relative contributions to variability that are, intuitively, 'persistent'. Further details of this technique are provided by Sims (1980, 1982) and Doan, Litterman and Sims (1984). As these references point out, however, there is an important disadvantage to the technique; the choice of the **S** matrix is not unique, so that a different ordering of the z variables will change **S**, thus altering the $\psi_{ij,k}^*$ coefficients and hence the impulse response functions and variance decompositions. The extent of these changes will depend upon the size of the contemporaneous correlations between the components of the $\mathbf{v}_t$ vector. Sims (1981) recommends that various orderings of the variables be tried, with the subsequent sets of impulse response functions and variance decompositions then being checked for robustness and consistency.

The variance decomposition methodology has generated a great deal of interest, and has been the subject of much detailed analysis and criticism. The references to Sims and his coworkers contain spirited defences of the technique, and Sims (1987) is the most recent exposition of its advantages. Critics include, in particular, Leamer (1985) and Cooley and LeRoy (1985), who draw attention to, amongst other things, the fact that VARs cannot be regarded as 'structural' in the traditional econometric sense. Their argument is that, unless prior predeterminedness or weak exogeneity assump-

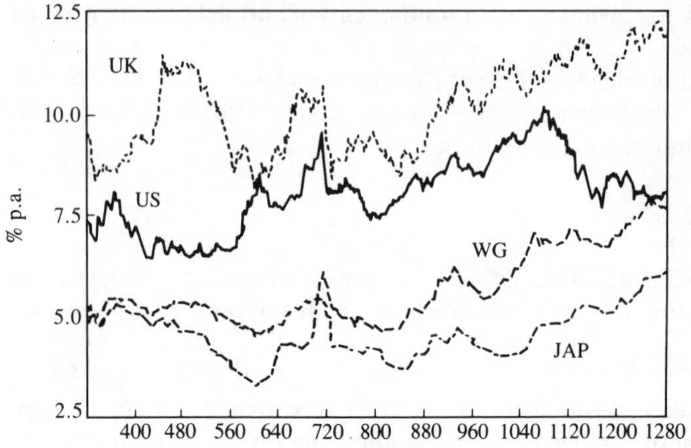

Figure 5.1 International bond yields (daily April 1986–December 1989)

tions concerning the presence or otherwise of contemporaneous variables in structural equations are made, innovations cannot be uniquely identified with a particular variable, thus invalidating the computed impulse response functions and variance decompositions. Such criticisms appear to imply that formal theoretical considerations should dictate the choice of variable ordering before triangularisation, although how different this is from the Sims recommendations above is debatable.

5.7 Vector ARMA models

A natural extension of the VAR is the vector ARMA process

$$\mathbf{z}_t = \sum_{i=1}^{p} \Phi_i \mathbf{z}_{t-i} + \mathbf{v}_t + \sum_{i=1}^{q} \Theta_i \mathbf{v}_{t-i},$$

or

$$\Phi(B)\mathbf{z}_t = \Theta(B)\mathbf{v}_t,$$

where

$$\Phi(B) = \mathbf{I} - \Phi_1 B - \ldots - \Phi_p B^p$$

and

$$\Theta(B) = \mathbf{I} + \Theta_1 B + \ldots + \Theta_q B^q,$$

which, of course, admits a VAR(∞) representation with $\pi(B) = \Theta^{-1}(B)\Phi(B)$. The presence of a vector MA component unfortunately complicates

Table 5.2 *Schwarz' criterion values and LR statistics for bond yield VARs*

j	$\mathcal{SC}(j)$	$\mathcal{LR}(j)$
1	32.11	233.0
2	32.06	168.5
3	32.03	133.6
4	32.02	114.4
5	32.01	88.0
6	31.99	62.9
7	31.97	33.6
8	31.95	

$\mathcal{LR}(j) \sim \chi^2_{4(8-j)}$

analysis somewhat. Details of such models, including estimation methods and model-building techniques, may be found in, for example, Mills (1990, chapter 14) and Lutkepohl (1991).

Example 5.6: Examining the international transmission of bond market movements

Figure 5.1 shows daily close-of-trade observations from 1 April 1986 to 29 December 1989, a total of 960 observations, on the redemption yields for government bonds with less than five years to maturity from four major bond markets, the US, the UK, West Germany (WG) and Japan. Although each of the series are non-stationary, a sequence of unit root tests reported in Mills and Mills (1991) confirms that each is an $I(1)$ process, so that the daily bond 'returns' can be regarded as being stationary and hence suitable for analysing in a VAR framework.

Table 5.2 presents $\mathcal{SC}$ values for lags $j=1,2,\ldots,m=8$, along with LR statistics for testing $m=8$ against $l=1,2,\ldots,7$. The minimum $\mathcal{SC}$ is found at $j=8$, and any shortening of the lag length produces a significant LR test statistic. We thus take $j=8$ as the order of our $g=4$-dimensional VAR, and summary statistics of the so estimated model are reported as table 5.3, along with an estimate of the contemporaneous correlation matrix of the residuals.

An interesting hypothesis to examine is whether movements in the US market have a tendency to lead (i.e., Granger-cause) movements in the other markets. Thus, defining $\mathbf{y}=$(UK, WG, Japan) and $\mathbf{x}=$(US), to use an obvious notation, feedback measures and associated Granger causality tests are computed using the methodology outlined in section 5.5.2, the

Table 5.3 *Summary statistics for the vector autoregression of example 5.6*

	VAR summary statistics			
	R^2	*s.e.*	$Q(20)$	
US	0.04	0.079	7.9	
UK	0.06	0.095	15.3	
WG	0.17	0.034	10.7	
Japan	0.14	0.037	15.3	
	Contemporaneous correlation matrix of VAR innovations			
	US	UK	WG	Japan
US	1			
UK	0.16	1		
WG	0.16	0.20	1	
Japan	0.06	0.13	0.14	1

Table 5.4 *Feedback measures and Granger-causality tests for the partition* y = *(UK, WG, Japan) and* x = *(US)*

	$\hat{F}$	$\mathscr{LR}(df)$
$x \rightarrow y$	0.1043	100.2(24)
$y \rightarrow x$	0.0403	38.7(24)
$y \cdot x$	0.0473	45.3(3)
y,x	0.1919	184.4(51)

results of which are reported in table 5.4. There is clear evidence of US bond yields Granger causing the other yields and of instantaneous causality, but there is somewhat less conclusive evidence of feedback from the other markets to the US, the estimate of $F_{\mathrm{y} \rightarrow \mathrm{x}}$ being significant at only the 0.029 level and a 95 per cent confidence interval being $(-0.0037, 0.0419)$.

Further analysis of the relationships between the markets can be undertaken through the computation of variance decompositions. From table 5.3, however, positive contemporaneous innovation correlations are observed, thus necessitating a prior ordering of the variables before impulse response functions and variance decompositions can be calculated.

Table 5.5 *Variance decompositions*

	Days ahead	Explained by US I	US II	UK I	UK II	WG I	WG II	Japan I	Japan II
US	1	100.0	95.6	0	2.4	0	1.7	0	0.3
	5	98.5	94.2	0.5	2.8	0.5	2.3	0.5	0.7
	10	97.3	92.9	0.8	3.1	1.2	2.9	0.7	1.1
	20	97.1	92.8	0.9	3.2	1.2	2.9	0.7	1.1
UK	1	2.6	0	97.4	98.3	0	0	0	1.7
	5	4.6	1.7	95.0	96.2	0.1	0.2	0.3	1.9
	10	5.1	2.2	93.6	94.6	0.6	0.9	0.7	2.3
	20	5.1	2.2	93.5	94.6	0.7	0.9	0.7	2.3
WG	1	2.7	0	3.1	3.4	94.2	94.6	0	2.0
	5	11.3	6.6	5.0	6.6	82.7	84.8	1.0	3.0
	10	13.4	8.3	4.9	6.5	81.7	82.9	1.4	3.6
	20	13.5	8.4	4.9	6.5	81.4	82.7	1.5	3.7
Japan	1	0.3	0	1.5	0	1.3	1.4	96.9	100.0
	5	2.3	1.3	2.9	1.4	2.1	1.1	92.7	96.2
	10	2.6	1.5	3.8	2.1	3.0	1.8	90.6	94.6
	20	2.7	1.6	3.9	2.2	3.2	1.9	90.2	94.2

Variance decompositions are presented in table 5.5 for two orderings; that based on the partition used for computing the feedback measures [I: (US, UK, WG, Japan)], and an ordering based upon the relative timing of the trading hours of the markets, [II: (Japan, UK, WG, US)]. Only the impulse response functions, which show the response of each variable to a unit innovation in the others, from ordering I are presented in table 5.6, as those obtained from ordering II were very similar at lags greater than zero.

The variance decompositions differ somewhat across the two orderings: as is to be expected, the influence of the US market is somewhat diminished in ordering II when it is at the 'bottom' rather than at the 'top', while the influence of Japan is slightly increased as its position in the ordering is correspondingly reversed. Nonetheless, certain general features do stand out. The US and UK are the most 'exogenous' of the markets, the West German is the most open to international influences, and Japan appears to act as a follower, reacting to innovations in other markets, rather than its own innovations having an influence on the others.

Table 5.6 *Impulse responses*

Day after shock	Response of US to unit innovations in			
	US	UK	WG	Japan
0	0.98	0	0	0
1	0.06	0.01	−0.10	0.05
2	−0.02	0.02	−0.14	0.07
3	0.09	−0.04	0.09	0.08
4	−0.02	−0.03	0.02	0.09
10	−0.03	−0.01	−0.02	−0.01
20	−0.00	−0.00	−0.10	−0.01
Day after shock	**Response of UK to unit innovations in**			
	US	UK	WG	Japan
0	0.19	0.97	0	0
1	0.16	0.07	0.01	−0.06
2	−0.01	−0.01	−0.05	0.09
3	0.06	0.04	0.06	0.05
4	0.05	−0.01	0.02	0.07
10	0.01	0.01	−0.04	−0.01
20	−0.00	0.00	−0.01	−0.00
Day after shock	**Response of WG to unit innovations in**			
	US	UK	WG	Japan
0	0.07	0.06	0.95	0
1	0.13	0.05	0.11	0.02
2	0.04	0.03	0.00	0.00
3	0.02	−0.00	0.00	0.01
4	0.01	0.00	0.00	0.09
10	0.01	0.01	−0.01	0.02
20	0.00	−0.00	0.00	0.00
Day after shock	**Response of Japan to unit innovations in**			
	US	UK	WG	Japan
0	0.03	0.05	0.12	0.97
1	0.06	0.02	0.07	0.04
2	0.02	0.02	−0.00	0.21
3	0.01	0.02	0.06	0.07
4	0.02	0.03	0.07	0.06
10	0.01	0.01	0.01	0.04
20	0.00	0.00	0.00	0.01

The impulse responses in table 5.6 show that shocks in one market are rapidly transmitted to other markets, for in all cases the innovation has virtually worked itself out within two days. This is also the case for each individual market's response to its own innovations. These quick reactions to innovations suggest that the behaviour of bond markets seem to be broadly consistent with the notion of informationally efficient international financial markets, implying that it would be difficult to earn unusual profits by operating in a particular market based on observed developments in other markets.

6 Regression techniques for integrated financial time series

Chapter 5 has developed regression techniques for modelling relationships between *non-integrated* time series: as we have seen in earlier chapters, however, many financial series are integrated, often able to be characterised as $I(1)$ processes, and the question thus arises as to whether the presence of integrated variables affects our standard regression results and conventional procedures of inference.

To this end, section 1 investigates this question through the analysis of spurious regressions between integrated time series. This leads naturally on to the concept of cointegration, which is introduced in section 2. Testing for cointegration is the material of sections 3 and 4, the latter placing cointegration within the more general VAR framework introduced in chapter 5.5. Alternative representations of cointegrated systems are the subject of section 5, with their estimation being considered in section 6.

6.1 Spurious regressions

We begin by considering the simulation example analysed by Granger and Newbold (1974) in an important article examining some of the likely empirical consequences of nonsense, or *spurious*, regressions in econometrics. They consider a situation in which y_t and x_t are generated by the *independent* random walks

$$y_t = y_{t-1} + v_t, \quad x_t = x_{t-1} + w_t, \quad t = 1,2,\ldots,$$

where v_t is iid $(0,\sigma_v^2)$ and w_t is iid $(0,\sigma_w^2)$. The regression of y_t on a constant and x_t is then considered

$$y_t = \hat{\alpha} + \hat{\beta}x_t + \hat{u}_t, \quad t = 1,2,\ldots,T. \tag{6.1}$$

With $T=50$, $y_0 = x_0 = 100$ and v_t and w_t drawn from independent $N(0,1)$ populations, Granger and Newbold report a rejection rate of 76 per cent when testing the (correct) null hypothesis that $\beta = 0$ in the regression (6.1) using the conventional t-statistic for assessing the significance of $\hat{\beta}$. Moreover, when five independent random walks are included as regressors in a multiple regression, the rejection rate of a conventional F-statistic, testing that the coefficient vector is zero rises to 96 per cent. For regressions involving independent ARIMA(0,1,1) series the corresponding rejection rates are 64 per cent and 90 per cent and Granger and Newbold thus conclude that conventional significance tests are seriously biased towards rejection of the null hypothesis of no relationship, and hence towards acceptance of a *spurious* relationship, when the series are generated as statistically independent integrated processes.

Moreover, such regression results are frequently accompanied by high R^2 values and highly autocorrelated residuals, as indicated by very low Durbin–Watson (dw) statistics. These findings led Granger and Newbold (1974) to suggest that, in the joint circumstances of a high R^2 and a low dw statistic (a useful rule of thumb being $R^2 > dw$), regressions should be run on the first differences of the variables. Further empirical evidence in favour of first differencing in regression models is provided by Granger and Newbold (1977, pp. 202–14) and Plosser and Schwert (1978).

These essentially empirical conclusions have since been given an analytical foundation by Phillips (1986), who makes much weaker assumptions about the innovations $\xi_t' = (v_t, w_t)'$ than those made above. By defining the *partial sum process*

$$S_t = \sum_{j=1}^{t} \xi_j, \quad S_0 = 0, \tag{6.2}$$

the set of conditions placed upon the innovation sequence $\{\xi_t\}_1^\infty$ can be defined as follows

(a) $E(\xi_t) = 0$ for all t.
(b) $\sup_{i,t} E|\xi_{it}|^{\beta+\epsilon} < \infty$ for some $\beta > 2$ and $\epsilon > 0$, $i = 1,2$ ($\xi_{1t} = v_t$, $\xi_{2t} = w_t$).
(c) $\Sigma = \lim_{T\to\infty} T^{-1} E(S_T S_T')$ exists and is positive definite.
(d) $\{\xi_t\}_1^\infty$ is strong mixing (see chapter 5.1.1).

These conditions permit y_t and x_t to be correlated $I(1)$ processes whose differences are said to be *weakly dependent* with possibly heterogeneously distributed innovations. This allows a wide variety of possible data-generating mechanisms, including, for example, the ARIMA(p,1,q) model. Condition (b) controls the allowable heterogeneity of the process, whereas (d) controls the extent of permissible temporal dependence in the process in relation to the probability of outlier occurrence: there is a tradeoff between a higher probability of outliers and the extent of temporal dependence. Condition (c) ensures that a suitably normalised function of S_t has a non-degenerate asymptotic distribution. Note that $\{\xi_t\}_1^\infty$ need not be (weakly) stationary, but if it is, condition (c) is automatically satisfied,

$$\Sigma = E(\xi_1\xi_1') + \sum_{k=1}^{\infty} E(\xi_1\xi_k' + \xi_k\xi_1'),$$

and the convergence of ξ_t is implied by the mixing condition (d). For univariate ξ_t, these conditions are those referred to in chapter 3.1.1 as being employed by Phillips (1987a, 1987b) in the context of testing for unit roots in individual series.

In the special case when v_t and w_t are independent

$$\Sigma = \begin{bmatrix} \sigma_v^2 & 0 \\ 0 & \sigma_w^2 \end{bmatrix},$$

where

$$\sigma_v^2 = \lim_{T\to\infty} T^{-1}E(P_T^2), \quad \sigma_w^2 = \lim_{T\to\infty} T^{-1}E(Q_T^2),$$

and

$$P_t = \sum_{j=1}^{t} v_j, \quad Q_t = \sum_{j=1}^{t} w_j, \quad P_0 = Q_0 = 0.$$

Phillips (1986) shows that, under these conditions, suitably standardised sample moments of the sequences $\{y_t\}_1^\infty$ and $\{x_t\}_1^\infty$ converge weakly to appropriately defined functionals of Brownian motion, rather than to constants as in the non-integrated regressor case discussed in chapter 5, which assumes that y_t and x_t are, for example, ergodic. Standard Brownian motion, which we may denote $W(r)$, is a stochastic process having the following property (see, for example, Hall and Heyde, 1980): for each $0 \le r_1 \le r_2 \le \ldots \le r_j \le 1$, $(W(r_1), W(r_2) - W(r_1), \ldots, W(r_j) - W(r_{j-1}))$ has a multivariate normal distribution with zero mean, zero covariances, and variances equal to $r_1, r_2 - r_1, \ldots, r_j - r_{j-1}$, i.e., it is a continuous stochastic process having independent, normally distributed increments. It can thus be regarded as the limit of a continuous time random walk as the step sizes become infinitesimally small.

A consequence of this result is that the standard distributional results of least squares regression, based as they are on ratios of sample moments converging to constants, break down. While not providing too great a level of rigour, a sketch of the derivation of this crucial result is nonetheless illuminating. We begin by noting that we may write $y_t = P_t + y_0$ and $x_t = Q_t + x_0$, where the initial conditions can either be constants or have certain specified distributions, from which we construct the standardised sums

$$Y_T(r) = T^{-\frac{1}{2}}\sigma_v^{-1}P_{[Tr]} = T^{-\frac{1}{2}}\sigma_v^{-1}P_{j-1},$$

$$X_T(r) = T^{-\frac{1}{2}}\sigma_w^{-1}Q_{[Tr]} = T^{-\frac{1}{2}}\sigma_w^{-1}Q_{j-1},$$
$$(j-1)/T \leq r < j/T, \quad j=1,\ldots,T,$$

where $[Tr]$ denotes the integer part of Tr, and

$$Y_T(1) = T^{-\frac{1}{2}}\sigma_v^{-1}P_T, \quad X_T(1) = T^{-\frac{1}{2}}\sigma_w^{-1}Q_T.$$

Using the more general partial sum process S_t in (6.2), we can also construct

$$Z_T(r) = T^{-\frac{1}{2}}\Sigma^{-\frac{1}{2}}S_{[Tr]} = T^{-\frac{1}{2}}\Sigma^{-\frac{1}{2}}S_{j-1},$$

and

$$Z_T(1) = T^{-\frac{1}{2}}\Sigma^{-\frac{1}{2}}S_T,$$

where $\Sigma^{\frac{1}{2}}$ is the positive definite square root of Σ. Phillips (1987c) proves that, as $T\uparrow\infty$, $Z_T(r)$ converges weakly to the vector Brownian motion $Z(r)$, i.e.

$$Z_T(r) \Rightarrow Z(r) \text{ on } C^2 = C[0,1] \times C[0,1],$$

where $C[0,1]$ is the space of all real-valued continuous functions on the interval $[0,1]$. From the properties of Brownian motion, $Z(r)$ is multivariate normal, with independent increments (so that $Z(s)$ is independent of $Z(r)-Z(s)$ for $0<s<r\leq 1$) and with independent elements (so that $Z_i(r)$ is independent of $Z_j(r)$, $i \neq j$).

When the sequences $\{v_t\}$ and $\{w_t\}$ are independent

$$Z_T(r) = \begin{bmatrix} Y_T(r) \\ X_T(r) \end{bmatrix}, \quad Z(r) = \begin{bmatrix} V(r) \\ W(r) \end{bmatrix},$$

and hence

$$X_T(r) \Rightarrow W(r), \quad Y_T(r) \Rightarrow V(r) \text{ as } T\uparrow\infty,$$

where $W(r)$ and $V(r)$ are independent Brownian motions on $C[0,1]$.

Now consider, for example, a suitably standardised first moment of x_t, viz.

$$T^{-3/2}\sum_{t=1}^{T} x_t = T^{-3/2}\sum_{t=1}^{T}(Q_{t-1} + w_t + x_0)$$
$$= \sigma_W T^{-1}\sum_{t=1}^{T}(T^{-1/2}\sigma_w^{-1})Q_{t-1} + T^{-3/2}\sum_{t=1}^{T}(w_t + x_0).$$

Phillips (1986, Mathematical Appendix *A.2*) proves that, as $T\uparrow\infty$, this weakly converges in distribution to that of the functional $\sigma_w\int_0^1 W(r)\mathrm{d}r$. Since $W(r)$ is Gaussian with mean zero and independent increments, the limiting distribution of $T^{-3/2}\sum_1^T x_t$ is thus normal with mean zero and variance given by

$$\sigma_w^2 E\left\{\int_0^1\int_0^1 W(r)W(s)\mathrm{d}r\mathrm{d}s\right\} = 2\sigma_w^2\int_0^1\int_0^r E\{W(r)W(s)\}\mathrm{d}s\mathrm{d}r$$

$$= 2\sigma_w^2\int_0^1\int_0^r s\mathrm{d}s\mathrm{d}r = \sigma_w^2/3 .$$

Similar results hold for other standardised moments of $\{y_t\}$ and $\{x_t\}$ and Phillips uses these to show that, as $T\uparrow\infty$

(i) $\hat{\beta}\Rightarrow(\sigma_v/\sigma_w)(\mu_{vw}/\mu_{ww})$,
(ii) $T^{-\frac{1}{2}}\hat{\alpha}\Rightarrow\sigma_v\phi$,
(iii) $T^{-\frac{1}{2}}t_\beta\Rightarrow\mu_{vw}/\nu^{\frac{1}{2}}$,
(iv) $T^{-\frac{1}{2}}t_\alpha\Rightarrow\phi\mu_{ww}[\nu\int_0^1 W(r)^2\mathrm{d}r]^{-\frac{1}{2}}$,
(v) $R^2\Rightarrow\mu_{vw}^2/\mu_{ww}\mu_{vv}$,
(vi) $dw \underset{p}{\rightarrow} 0$.

In these results, we use the functionals

$$\mu_{ab} = \int_0^1 a(r)b(r)\mathrm{d}r - \int_0^1 a(r)\mathrm{d}r\int_0^1 b(r)\mathrm{d}r$$

$$\phi = \int_0^1 V(r)\mathrm{d}r - \frac{\mu_{vw}}{\mu_{ww}}\int_0^1 W(r)\mathrm{d}r ,$$

and

$$\nu = \mu_{vv}\mu_{ww} - \mu_{vw} .$$

As Phillips (1986) remarks, these analytical results go a long way towards explaining the Monte Carlo findings reported by Granger and Newbold (1974). Results (iii) and (iv) show that the conventional t-ratios, t_α and t_β, do not have limiting distributions, in fact diverging as $T\uparrow\infty$, so that there are *no* asymptotically correct critical values for these tests. We should thus expect the rejection rate when these tests are based on a critical value delivered from conventional asymptotics (such as 1.96) to continue to increase with sample size, and this is consistent with the findings of Granger and Newbold.

Results (i) and (ii) show that, in contrast to the usual results of regression theory, $\hat{\alpha}$ and $\hat{\beta}$ do not converge in probability to constants as $T\uparrow\infty$: indeed, $\hat{\beta}$ has a non-degenerate limiting distribution and the distribution of $\hat{\alpha}$ actually diverges. Thus the uncertainty about the regression (6.1) stemming from its spurious nature persists asymptotically in these limiting distributions, this being a consequence of the sample moments of y_t and x_t (and their joint sample moments) not converging to constants but, upon appropriate standardisation, converging weakly to random variables.

Results (v) and (vi) show that R^2 has a non-degenerate limiting distribution and that *dw* converges in probability to zero as $T\uparrow\infty$. Low values for *dw* and moderate values of R^2 are therefore to be expected in spurious regressions such as (6.1) with data generated by integrated processes, again confirming the simulation findings reported by Granger and Newbold.

These results are easily extended to multiple regressions of the form

$$y_t=\hat{\alpha}+\hat{\beta}'\mathbf{x}_t+\hat{u}_t, \quad t=1,2,\ldots,T, \tag{6.3}$$

where $\mathbf{x}_t=(x_{1t},\ldots,x_{kt})$ is a vector $I(1)$ process. Phillips (1986) shows that analogous results to (i)–(vi) above hold for (6.3) and, in particular, that the distribution of the customary regression *F*-statistic for testing the joint significance of the vector $\hat{\beta}$ diverges as $T\uparrow\infty$ and so there are no asymptotically correct critical values for this statistic either. Moreover, the divergence rate for the *F*-statistic is $O(T)$, which is greater than the divergence rate of $O(T^{\frac{1}{2}})$ for the individual *t*-tests. In a regression with many regressors, therefore, we might expect a noticeably greater rejection rate for the block *F*-test than for the individual *t*-tests or for a test with fewer regressors, and this is again consistent with the results reported by Granger and Newbold.

We should emphasise that, although the derivation of the asymptotic results has assumed independence of y_t and $\mathbf{x}_t$, so that the true values of α and β are zero, this is not crucial to the major conclusions. Although the correlation properties of the time series do have quantitative effects on the limiting distributions, these being introduced via the parameters of the limiting covariance matrix Σ in the bivariate regression analysed in detail above, such effects do not interfere with the main qualitative results: viz. that $\hat{\alpha}$ and $\hat{\beta}$ do not converge in probability to constants, that the distributions of *F*- and *t*-statistics diverge as $T\uparrow\infty$, and that *dw* converges in probability to zero whereas R^2 has a non-degenerate limiting distribution as $T\uparrow\infty$.

It should also be emphasised that, in the general set up discussed here, where both y_t and $\mathbf{x}_t$ are $I(1)$ processes, the error, u_t, since it is by definition a linear combination of $I(1)$ processes, will also be integrated, unless a special restriction, to be discussed subsequently, holds. Moreover, the usual respecification of the model to include y_{t-1} as an additional regressor on the finding of a very low *dw* value will have pronounced consequences: the estimated coefficient on y_{t-1} will converge to one, while that on the integrated regressor will converge to zero, thus highlighting the spurious nature of the static regression.

Indeed, the spurious nature of the regression is, in fact, a consequence of the error being $I(1)$. Achieving a stationary, or $I(0)$, error is usually a minimum criterion to meet in econometric modelling, for much of the focus of recent developments in the construction of dynamic regression models has been to ensure that the error is not only $I(0)$ but white noise. Whether the error in a regression between integrated variables is stationary is thus a matter of considerable importance.

6.2 Cointegrated processes

As we have just remarked, a linear combination of $I(1)$ processes will usually also be $I(1)$. In general, if x_t and y_t are both $I(d)$, then the linear combination

$$u_t = y_t - ax_t \tag{6.4}$$

will usually be $I(d)$. It is possible, however, that u_t may be integrated of a lower order, say $I(d-b)$, where $b>0$, in which case a special constraint operates on the long-run components of the two series. If $d=b=1$, so that x_t and y_t are both $I(1)$ and dominated by 'long wave' components, u_t will be $I(0)$, and hence will not have such components: y_t and ax_t must therefore have long-run components that cancel out to produce u_t. In such circumstances, x_t and y_t are said to be *cointegrated*; we emphasise that it will *not* generally be true that there will exist such an a which makes $u_t \sim I(0)$ or, in general, $I(d-b)$.

A formal definition of cointegration is provided by Engle and Granger (1987): the components of the n-dimensional vector $\mathbf{z}_t$ are said to be *cointegrated of order d, b*, denoted $\mathbf{z}_t \sim CI(d,b)$, if (i) all components of $\mathbf{z}_t$ are $I(d)$; and (ii) there exists at least one vector $\alpha(\neq 0)$ such that $\mathbf{u}_t = \alpha'\mathbf{z}_t \sim I(d-b), b>0$. The vector α is called the *cointegrating vector*. There may be m cointegrating vectors, with $m<n$. When $1<m<n$, the cointegrating vectors may be denoted $\alpha_1, \ldots, \alpha_m$ and may be gathered together into the $n \times m$ matrix $\mathbf{B} = (\alpha_1, \ldots, \alpha_m)$. By construction, the rank of $\mathbf{B}$ is m, which is termed the *cointegrating rank* of $\mathbf{z}_t$. In the following discussion, we shall concentrate, as is customary, on the $d=b=1$ case.

The idea of cointegration can be related to the concept of *long-run equilibrium*, which we may illustrate by the bivariate relationship

$$y_t = ax_t,$$

or

$$y_t - ax_t = 0.$$

Thus u_t given by (6.4) measures the extent to which the 'system' is out of equilibrium, and can therefore be termed the 'equilibrium error'. Hence, since x_t and y_t are both $I(1)$, the equilibrium error will be $I(0)$ and u_t will rarely drift far from zero, and will often cross the zero line. In other words, equilibrium will occasionally occur, at least to a close approximation, whereas if x_t and y_t are not cointegrated, so that $u_t \sim I(1)$, the equilibrium error will wander widely and zero-crossings would be very rare, suggesting that under such circumstances the concept of equilibrium has no practical implications.

It is this feature of cointegration that links it with the analysis of spurious regressions. Condition (c) on the innovation sequence $\{\xi_t\}$ requires that the limiting covariance matrix Σ be non-singular. If we allow Σ to be singular, then the asymptotic theory yielding the results (i)–(vi) no longer holds. In general, we have

$$\Sigma = \begin{bmatrix} \sigma_v^2 & \sigma_{vw} \\ \sigma_{vw} & \sigma_w^2 \end{bmatrix},$$

so that for Σ to be singular, $|\Sigma| = \sigma_v^2\sigma_w^2 - \sigma_{vw}^2 = 0$. This implies that $\Sigma\gamma = 0$, where $\gamma' = (1, -a)$ and $a = \sigma_{vw}/\sigma_w^2$. Singularity of Σ is a necessary condition for y_t and x_t to be cointegrated (Phillips, 1986, Phillips and Ouliaris, 1990), since in this case $|\Sigma| = 0$ implies that the 'long-run' correlation between the innovations v_t and w_t, given by $\rho = \sigma_{vw}/\sigma_v\sigma_w$, is unity. For values of ρ less than unity, y_t and x_t are not cointegrated, and when $\rho = 0$, so that v_t and w_t are independent, we have Granger and Newbold's (1974) spurious regression.

What differences to the asymptotic regression theory for integrated processes result when y_t is cointegrated with a vector of regressors $\mathbf{x}_t$? Note that the equilibrium error u_t can be regarded as the error term in the regression of y_t on $\mathbf{x}_t$, i.e., we may consider the model

$$y_t = \alpha + \beta \mathbf{x}_t + u_t, \quad t = 1, 2, \dots, \tag{6.5a}$$

where

$$\mathbf{x}_t = \pi + \mathbf{x}_{t-1} + w_t, \quad t = 1, 2, \dots \tag{6.5b}$$

Unlike the case where the variables are not cointegrated, $\{u_t\}_1^\infty$ is now known to be $I(0)$, and hence we may allow the sequence $\{e_t'\}_1^\infty = \{(u_t, w_t)\}_1^\infty$ of joint innovations to satisfy conditions (a)–(d). A wide range of serial dependence and simultaneity, as well as non-stationarity, is then permitted in the system (6.5), where we note that drift ($\pi \neq 0$) may be allowed in the regressors. For example, none of the common exogeneity conditions need necessarily apply since contemporaneous correlations of the form $E(\mathbf{x}_t u_t) \neq 0$ are allowed, and (6.5b) may be regarded as the reduced form of a simultaneous equations system in which the exogenous variables are driven by a quite general vector $I(1)$ process.

Given these assumptions on the joint innovation process $\{e_t\}$, Park and Phillips (1988) (see also Phillips and Durlauf, 1986, and Stock, 1987) prove that $\hat{\alpha}$ and $\hat{\beta}$ are now consistent, and indeed converge at a faster rate ($O(T^{-1})$) than in conventional regression theory. An important consequence of this result is that there is no asymptotic simultaneous equations or measurement error bias in regressions such as (6.5a) when the regressors form an integrated process: for an intuitive interpretation of this feature of cointegrating regressions, see the discussions in Phillips and Durlauf (1986, p. 482) and Stock and Watson (1988a).

Although consistent, $\hat{\alpha}$ and $\hat{\beta}$ are not, however, asymptotically normal when appropriately centred and scaled since, as in the non-cointegrated case, suitably scaled sample moments converge weakly to random rather than constant matrices, so that the asymptotic distributions are the functionals of Brownian motion discussed in section 6.1. As a consequence, standard tests of the type developed in chapter 5.1.2 no longer yield asymptotically distributed χ^2 criteria. When the regressor vector $\mathbf{x}_t$ is without drift ($\pi = 0$), Park and Phillips (1988) construct modified test statistics in which the estimate of the contemporaneous covariance matrix of $\{e_t\}$, Σ_e say, that is used in these statistics is replaced by a consistent estimate of the 'long-run' covariance matrix

$$\Omega_e = \Sigma_e + \Lambda_e + \Lambda_e',$$

where

$$\Lambda_e = \sum_{i=2}^{\infty} E(e_1 e_i').$$

Consistent estimation of Ω_e is discussed in Phillips and Durlauf (1986). If $\{u_t\}$ is white noise, however, standard testing procedures can be used.

If the regressors contain drifts, the asymptotic covariance matrix of $\hat{\beta}$ will be singular since the regressors will be perfectly correlated asymptotically. This is because an $I(1)$ variable with drift can always be expressed as the sum of a time trend and an $I(1)$ variable without drift, i.e.

$$\nabla x_t = \pi + w_t = x_0 + \pi t + \nabla \tilde{x}_t, \qquad \nabla \tilde{x}_t = w_t,$$

so that the correlation between two such variables will be dominated by the trend, which will be $O(T)$, compared to the driftless $I(1)$ components, which will only be $O(T^{\frac{1}{2}})$. This suggests that these variables should be detrended and a time trend added to (6.5a). The estimator of the coefficient of the trend will be asymptotically normal, while the estimators of the coefficients on their 'driftless' components will have the non-standard distribution discussed above.

A special case is when $\mathbf{x}_t$ is a scalar and has non-zero drift ($\pi \neq 0$). In this case, $T^{\frac{1}{2}}(\hat{\alpha} - \alpha)$ and $T^{3/2}(\hat{\beta} - \beta)$ are asymptotically normal, with covariance matrices that can be calculated using the standard formulae and procedures developed in chapter 5 (see West, 1988a). That this result only holds when there is just one integrated regressor may be explained by noting that the non-zero drift π imparts a trend into the regression. It is the trend coefficient, $\pi\beta$, that is asymptotically normal and this allows the result on $\hat{\beta}$ to follow. When there are two or more integrated regressors with drift, the trend coefficient becomes a linear combination of the different drifts, and only this combination can be identified and is asymptotically normal. The vector $\hat{\beta}$ can only be estimated by the coefficients on driftless $I(1)$ regressors and this will have the non-standard asymptotic distribution.

Another special case is when the regressors in (6.5) are *strictly exogenous*, i.e., when $\{u_t\}$ and $\{w_t\}$ are generated independently of each other. In such circumstances, Park and Phillips (1988) show that $\hat{\alpha}$ and $\hat{\beta}$ are again asymptotically normal on appropriate standardisation.

If a time trend is included as an additional regressor in (6.5a) then Park and Phillips (1988) show that the asymptotic results for the LS estimators of α and β remain valid, although the estimator of the coefficient on the time trend depends on π. Furthermore, if additional stationary regressors are included in (6.5a) then their coefficients will be asymptotically normal.

6.3 Testing for cointegration in regressions

Given the crucial role that cointegration plays in regression models with integrated variables, it is important to be able to test for its presence. A number of tests have been proposed that are based on the residuals from the *cointegrating regression*:

$$\hat{u}_t = y_t - \hat{\alpha} - \hat{\beta}\mathbf{x}_t.$$

Such residual-based procedures seek to test a null hypothesis of *no* cointegration by using unit root tests applied to $\hat{u}_t$. Perhaps the simplest test to use is the usual Durbin–Watson *dw* statistic but, since the null is that $\hat{u}_t$ is $I(1)$, the value of the test statistic under this null is $dw=0$, with rejection in favour of the $I(0)$ alternative occurring for values of *dw greater* than zero (Sargan and Bhargava, 1983, Bhargava, 1986). As is well known, the conventional critical values of the *dw* statistic depend upon the underlying processes generating the observed data, and Engle and Granger (1987) and Phillips and Ouliaris (1988) provide critical values, for various sample sizes and generating processes, in the 'non-standard' case considered here.

Engle and Granger (1987) prefer to use the t-ratio on $\hat{u}_{t-1}$ from the regression of $\nabla \hat{u}_t$ on $\hat{u}_{t-1}$ and lagged values of $\nabla \hat{u}_t$, in a manner analogous to the unit root testing approach for an observed series discussed in chapter 3 (see, for example, equation (3.3)). The problem here is that, since $\hat{u}_t$ is derived as a residual from a regression in which the cointegrating vector is estimated, and since if the null of non-cointegration was true such a vector would not be identified, using the τ_μ critical values would reject the null too often, because least squares will seek the cointegrating vector which minimises the residual variance and hence is most likely to result in a stationary residual series.

Engle and Granger (1987) and Engle and Yoo (1987) provide critical values of the appropriate distribution, which we shall denote $\hat{\tau}_\mu$, obtained by Monte Carlo simulation for various data generating processes, sample sizes and numbers of regressors. Phillips and Ouliaris (1990) obtain the limiting asymptotic distribution of $\hat{\tau}_\mu$ and provide critical values for up to five regressors when the cointegrating regression contains no constant, just a constant, and both a constant and a time trend. These critical values must, however, be used carefully, and Hansen (1992) proposes some modifications when a time trend is present in the cointegrating regression. A more complete set of critical values, calculated using response surface estimation, is provided by MacKinnon (1990); $\hat{\tau}_\mu$ tests of cointegration are now programmed as specific commands in both *MICROTSP 7.0* and *MICROFIT 3.0*, with the MacKinnon critical values being reported in both.

Phillips and Ouliaris (1990) also present critical values of the limiting distribution of the $Z(\tau_\mu)$ statistic, the t-ratio from the regression of ∇u_t on u_{t-1} with standard error adjusted as in equation (3.5), as well as a third statistic based on the coefficient estimate from this regression, and two further statistics, a 'variance ratio test' and a 'multivariate trace statistic', both of which can be obtained, at some extra computational cost, from the cointegrating regression. Only limited evidence has accumulated on the finite sample performance of the various statistics: Engle and Granger (1987) express a preference for the $\hat{\tau}_\mu$ statistic, and this certainly has the advantage of being simple to compute.

6.4 Testing for cointegration using VARs

Rather than work solely within a regression formulation, cointegration can also be examined within a VAR framework. Phillips and Durlauf (1986) consider the n-dimensional vector $\{\mathbf{z}_t\}_1^\infty$ to be generated by the VAR(1) process

$$\mathbf{z}_t = \mathbf{A}\mathbf{z}_{t-1} + \mathbf{u}_t, \tag{6.6a}$$

$$\mathbf{A} = \mathbf{I}_n, \tag{6.6b}$$

where $\{\mathbf{u}_t\}_1^\infty$ satisfies conditions (a)–(d) of section 1 and where the initial condition $\mathbf{z}_0$ can be either a constant or a random variable having a certain specified distribution independent of T, the sample size. It is convenient to allow $\{\mathbf{u}_t\}_1^\infty$ to be weakly stationary so that the 'long-run' error covariance matrix Σ is given by

$$\Sigma = E(\mathbf{u}_1\mathbf{u}_1') + \sum_{k=1}^{\infty} E(\mathbf{u}_1\mathbf{u}_k' + \mathbf{u}_k\mathbf{u}_1') = \Sigma_0 + \Sigma_1 + \Sigma_1' = 2\pi f_{\mathbf{uu}}(0), \tag{6.7}$$

where $f_{\mathbf{uu}}(\lambda)$ is the spectral density matrix of $\mathbf{u}_t$ at frequency λ, so that $f_{\mathbf{uu}}(0)$ is the spectral density at the origin.

Phillips and Durlauf show that (6.6) allows for quite general vector ARMA specifications such as $(1-B)\mathbf{A}(B)\mathbf{z}_t = \Theta(B)\mathbf{e}_t$ and go on to develop an asymptotic theory for VARs in the presence of unit roots. Defining

$$\mathbf{Z}' = [\mathbf{z}_1, \ldots, \mathbf{z}_T] \text{ and } \mathbf{Z}'_{-1} = [\mathbf{z}_0, \ldots, \mathbf{z}_{T-1}],$$

the matrix of regression coefficients from a (multivariate) regression of $\mathbf{z}_t$ on $\mathbf{z}_{t-1}$ is given by

$$\hat{\mathbf{A}} = \mathbf{Z}'\mathbf{Z}'_{-1}(\mathbf{Z}'_{-1}\mathbf{Z}_{-1})^{-1},$$

while the estimate of the associated error covariance matrix Σ_0 is

$$\hat{\Sigma}_0 = T^{-1}\mathbf{Z}'(\mathbf{I} - \mathbf{P}_{\mathbf{Z}_{-1}})\mathbf{Z},$$

where $\mathbf{P}_\mathbf{D} = \mathbf{D}(\mathbf{D}'\mathbf{D})^{-1}\mathbf{D}'$ for any matrix $\mathbf{D}$ of full column rank. They prove that $\hat{\mathbf{A}}$ is consistent, i.e., $\hat{\mathbf{A}} \xrightarrow{p} \mathbf{I}_n$, even in the presence of substantial serial correlation, and its asymptotic distribution, although non-normal, has the same form for a wide variety of different error processes, thus permitting heterogeneous error variances as well as weak dependence in the errors. Furthermore, $\hat{\Sigma}_0$ is also consistent and, unlike $\hat{\mathbf{A}}$, asymptotically normal: this is explained by the fact that, since $\hat{\mathbf{A}} \xrightarrow{p} \mathbf{I}_n$, the residuals from the regression are asymptotically stationary and thus, in effect, provide consistent estimates of the innovation process $\mathbf{u}_t$.

The vector $\mathbf{z}_t$ will be cointegrated if there exists a vector α for which $\mathbf{v}_t = \alpha'\mathbf{z}_t$ is weakly stationary. It follows directly from (6.6) that

$$\alpha'\mathbf{u}_t = \mathbf{v}_t - \mathbf{v}_{t-1},$$

so that some combination of the innovations in (6.6) has an MA representation with a unit root. Phillips and Ouliaris (1988) then deduce that the system (6.6) is cointegrated (i) *only if*

$$\alpha'\Sigma\alpha = 0, \tag{6.8}$$

and (ii) *if and only if*

$$\alpha' f_{\mathbf{uu}}(\lambda)\alpha = c\lambda^2 + o(\lambda^2) \text{ as } \lambda \to 0, \tag{6.9}$$

for some constant c (possibly zero). The necessary condition (6.8) implies that Σ is singular. When there are several distinct cointegrating vectors $\alpha_i (i = 1, \ldots, m < n)$, $\Sigma\alpha_i = 0$ (or $\Sigma\mathbf{B} = 0$) and Σ has m zero latent roots.

Condition (6.9) is both necessary and sufficient: it tells us that $\alpha' f_{\mathbf{uu}}(\lambda)\alpha$ is not only zero at the origin $\lambda=0$ but flat as well.

The singularity of Σ invalidates the limiting distribution theory for $\hat{\mathbf{A}}$ and $\hat{\Sigma}_0$, the requisite theory being provided by Phillips and Ouliaris (1988, section 3). They also develop tests of cointegration based on the latent roots of a consistent estimate of the covariance matrix Σ given in (6.7), this being provided by an estimate of the smoothed periodogram of $\mathbf{u}_t = \nabla \mathbf{z}_t$ computed using a rectangular spectral window.

Denoting this estimate as $\hat{\Sigma}$, we need to assess whether the smallest roots of $\hat{\Sigma}$ are negligible or, in other words, statistically insignificant. Assuming that the latent roots of Σ, λ_i $(i=1,\ldots,n)$, are distinct and are ordered such that $\lambda_1>\lambda_2>\ldots>\lambda_n$, then the corresponding roots of $\hat{\Sigma}$, $l_1\geq l_2\geq\ldots\geq l_n$, will be asymptotically independent of each other. The standardised variates $\kappa^{\frac{1}{2}}(l_i-\lambda_i)/\lambda_i (i=1,\ldots,n)$, where κ is the number of ordinates used in the estimation of the smoothed periodogram, are then asymptotically $(\kappa\uparrow\infty)$ $N(0,1)$, i.e., $l_i\sim N(\lambda_i,\lambda_i/\kappa)$. Confidence bounds for the smallest latent root λ_n and for the sum of the s smallest latent roots of Σ can then be constructed using standard methods.

As Phillips and Ouliaris (1988) point out, the latent roots will depend on the units of measurement of the variables that comprise $\mathbf{z}_t$. In such circumstances it may be better to work with dimensionless quantities, which may be constructed by defining the matrix

$$\mathbf{P}=\Sigma_0^{-1/2}\Sigma\Sigma_0^{-1/2}=\mathbf{I}+\sum_{k=1}^{\infty}(\Gamma_k+\Gamma_k'),$$

where

$$\Gamma_k=\Sigma_0^{-1/2}E(\mathbf{u}_0\mathbf{u}_k')\Sigma_0^{-1/2}$$

and $\Sigma_0^{1/2}$ is the positive definite square root of Σ_0. $\mathbf{P}$ can be estimated as

$$\hat{\mathbf{P}}=\hat{\Sigma}_0^{-1/2}\hat{\Sigma}\hat{\Sigma}_0^{-1/2}.$$

Defining the latent roots of $\mathbf{P}$ and $\hat{\mathbf{P}}$ to be $\rho_1>\rho_2>\ldots>\rho_n$ and $r_1\geq r_2\geq\ldots\geq r_n$, respectively, then in the same way as before, the latent roots r_i are (approximately) independently distributed as $N(\rho_i,\rho_i^2/\kappa)$ for large κ. Thus, for the smallest root ρ_n of $\mathbf{P}$, we have the upper and lower $100(1-\alpha)$ per cent confidence limits:

$$\rho_n\leq r_n+r_n z_\alpha/\kappa^{\frac{1}{2}}, \tag{6.10}$$

and

$$r_n-r_n z_\alpha/\kappa^{\frac{1}{2}}\leq\rho_n, \tag{6.11}$$

where z_α is the α percentage point of $N(0,1)$. Similarly, upper and lower

$100(1-\alpha)$ per cent confidence bounds for the ratio of the sum of the s smallest roots to the sum of all the roots of $\mathbf{P}$ are given by

$$\frac{\sum_{j=n-s}^{n} \rho_j}{\sum_{j=1}^{n} \rho_j} \leq \frac{\sum_{j=n-s}^{n} r_j}{\sum_{j=1}^{n} r_j} + z_\alpha D/\kappa^{\frac{1}{2}}, \tag{6.12}$$

and

$$\frac{\sum_{j=n-s}^{n} r_j}{\sum_{j=1}^{n} r_j} - z_\alpha D/\kappa^{\frac{1}{2}} \leq \frac{\sum_{j=n-s}^{n} \rho_j}{\sum_{j=1}^{n} \rho_j}, \tag{6.13}$$

where

$$D=\left[\left(\sum_{j=n-s}^{n} r_j\right)^2\left(\sum_{j=1}^{n-s-1} r_j^2\right)+\left(\sum_{j=1}^{n} r_j\right)^2\left(\sum_{j=n-s}^{n} r_j^2\right)\right]^{\frac{1}{2}}\bigg/\left(\sum_{j=1}^{n} r_j\right)^2.$$

After Monte Carlo simulation, Phillips and Ouliaris (1988) express a preference for using the unit-free ratio tests (6.12) and (6.13) and suggest the following decision rule: (a) reject the null hypothesis of no cointegration if the *upper* bound is less than 0.10, and (b) accept the null hypothesis if the *lower* bound is greater than 0.10.

Just as in the case of the *dw* test in conventional regression theory, this bounds test involves a region in which the test is inconclusive, arising when the 10 per cent decision rule lies between the computed lower and upper bounds for the minimum latent root. To circumvent this, Phillips and Ouliaris (1988, table 5) present certain critical values for various choices of n obtained by averaging the upper and lower percentile values of the null and alternative distributions, respectively, across some fifty data generating processes for the innovations.

When working within this VAR framework, two related approaches to testing for cointegration have also been developed. Rather than work with the VAR(1) system (6.6) with weakly dependent innovations, Johansen (1988, 1991, see also Johansen and Juselius, 1990) explicitly parameterises the system as a VAR(p) process

$$\mathbf{z}_t = \sum_{i=1}^{p} \mathbf{A}_i \mathbf{z}_{t-i} + \mathbf{e}_t, \tag{6.14}$$

where the innovations $\mathbf{e}_t$ are assumed to be *NID*. This can be rewritten as

$$\nabla \mathbf{z}_t = \sum_{i=1}^{p-1} \mathbf{C}_i \nabla \mathbf{z}_{t-i} + \mathbf{C}_p \mathbf{z}_{t-p} + \mathbf{e}_t, \tag{6.15}$$

where

$$\mathbf{C}_i = -\mathbf{I} + \sum_{j=1}^{i} \mathbf{A}_j, \quad i = 1, \ldots, p.$$

If there are m cointegrating vectors then *Granger's Representation Theorem* (Engle and Granger, 1987, Hylleberg and Mizon, 1989b) asserts that

$$\mathbf{C}_p = -\mathbf{I} + \mathbf{A}_1 + \ldots + \mathbf{A}_p = \mathbf{A}(1) = \gamma \mathbf{B}',$$

where γ and $\mathbf{B}$ are $n \times m$ matrices of rank m, $\mathbf{B}$ being the matrix of cointegrating vectors defined previously. Thus $\mathbf{C}_p$ is also of rank m. An LR test of the null hypothesis that there are at most m cointegrating vectors has been proposed by Johansen (1988) and is based on estimating (6.15) with the restriction $\mathbf{C}_p = \gamma \mathbf{B}'$ first imposed and then with it removed. The test statistic may be formed from the residuals from two auxiliary regressions

$$\nabla \mathbf{z}_t = \sum_{i=1}^{p-1} \Gamma_{0i} \nabla \mathbf{z}_{t-i} + \mathbf{w}_{0t}$$

and

$$\mathbf{z}_{t-p} = \sum_{i=1}^{p-1} \Gamma_{1i} \nabla \mathbf{z}_{t-i} + \mathbf{w}_{1t}.$$

Calculating the residual sample second-moment matrices

$$\hat{S}_{ij} = T^{-1} \sum_{t=1}^{T} \hat{\mathbf{w}}_{it} \hat{\mathbf{w}}'_{jt}, \quad i,j = 0,1,$$

the LR test of there being at most m cointegrating vectors takes the form

$$-2\ln Q_m = -T \sum_{i=m+1}^{n} \ln(1 - \hat{\lambda}_i),$$

where $\hat{\lambda}_{m+1}, \ldots, \hat{\lambda}_n$ are the $n-m$ smallest eigenvalues of $\hat{S}_{10}\hat{S}_{00}^{-1}\hat{S}_{01}$ with respect to $\hat{S}_{11}$. Under the null of at most m cointegrating vectors, or equivalently $n-m$ unit roots, the asymptotic distribution of this test statistic is given by

$$-2\ln Q_m \sim \mathrm{tr}\left[\int_0^1 B\,\mathrm{d}B' \left(\int_0^1 B(u)B(u)'\,\mathrm{d}u\right)^{-1} \int_0^1 \mathrm{d}B\,B'\right],$$

where B denotes $n-m$ dimensional vector Brownian motion. Various simulated quantiles of this distribution are tabulated in Johansen (1988), with extensions to incorporate drifts and seasonality tabulated in Johansen and Juselius (1990). This testing procedure is now incorporated in *MICROFIT 3.0*.

Rather than work with a VAR(p) representation for the levels $\mathbf{z}_t$, Stock and Watson (1988b) consider the vector MA representation of the differences $\nabla \mathbf{z}_t$

$$\nabla \mathbf{z}_t = \mathbf{e}_t + \mathbf{D}_1\mathbf{e}_{t-1} + \mathbf{D}_2\mathbf{e}_{t-2} + \ldots = \mathbf{D}(B)\mathbf{e}_t, \tag{6.16}$$

where $\mathbf{D}(B) = \mathbf{I} + \mathbf{D}_1 B + \mathbf{D}_2 B^2 + \ldots$ is an $n \times n$ matrix of lag polynomials. From Engle and Granger (1987), if $\mathbf{z}_t$ is cointegrated with m cointegrating vectors, the rank of $\mathbf{D}(1)$ is $n-m$ and $\mathbf{B}'\mathbf{D}(1) = 0$.

Recursive substitution in (6.16) yields

$$\mathbf{z}_t = \mathbf{z}_0 + \mathbf{D}(B)\sum_{s=0}^{t} \mathbf{e}_s . \tag{6.17}$$

Noting that $\mathbf{D}(B)$ can always be written as

$$\mathbf{D}(B) = \mathbf{D}(1) + (1-B)\mathbf{D}^*(B),$$

where $\mathbf{D}^*(B) = \mathbf{D}_0^* + \mathbf{D}_1^* B + \mathbf{D}_2^* B^2 + \ldots$ with $\mathbf{D}_j^* = -\sum_{j+1}^{\infty} \mathbf{D}_i$, $(j = 0,1,2,\ldots)$, equation (6.17) can be written as

$$\begin{aligned} \mathbf{z}_t &= \mathbf{z}_0 + \mathbf{D}(1)\sum_{s=0}^{t} \mathbf{e}_s + \mathbf{D}^*(B)\mathbf{e}_t \\ &= \mathbf{z}_0 + \mathbf{D}(1)\mathbf{E}_t + \mathbf{D}^*(B)\mathbf{e}_t , \end{aligned} \tag{6.18}$$

where $\mathbf{E}_t = \sum_1^t \mathbf{e}_s$. Since $\mathbf{D}(1)$ has rank $n-m$, there is a $n \times m$ matrix $\mathbf{H}_1$ (with rank m) such that $\mathbf{D}(1)\mathbf{H}_1 = 0$. Furthermore, if $\mathbf{H}_2$ is a $n \times (n-m)$ matrix with rank $n-m$ and with columns orthogonal to the columns of $\mathbf{H}_1$, then $\mathbf{F} = \mathbf{D}(1)\mathbf{H}_2$ has rank $n-m$. The $n \times n$ matrix $\mathbf{H} = (\mathbf{H}_1 : \mathbf{H}_2)$ is non-singular and $\mathbf{D}(1)\mathbf{H} = (\mathbf{0} : \mathbf{F}) = \mathbf{F}\mathbf{S}_{n-m}$, where $\mathbf{S}_{n-m}$ is the $(n-m) \times n$ selection matrix $(\mathbf{0} : \mathbf{I}_{n-m})$. Thus (6.18) can be rewritten

$$\begin{aligned} \mathbf{z}_t &= \mathbf{z}_0 + \mathbf{D}(1)\mathbf{H}\mathbf{H}^{-1}\mathbf{E}_t + \mathbf{D}^*(B)\mathbf{e}_t \\ &= \mathbf{z}_0 + \mathbf{F}\tau_t + \mathbf{a}_t , \end{aligned} \tag{6.19}$$

where $\tau_t = \mathbf{S}_{n-m}\mathbf{H}^{-1}\mathbf{E}_t$ and $\mathbf{a}_t = \mathbf{D}^*(B)\mathbf{e}_t$. Note that τ_t can be written as

$$\tau_t = \mathbf{S}_{n-m}\mathbf{H}^{-1}(\mathbf{E}_{t-1} + \mathbf{e}_t),$$

i.e.

$$\tau_t = \tau_{t-1} + \nu_t ,$$

where $\nu_t = \mathbf{S}_{n-m}\mathbf{H}^{-1}\mathbf{e}_t$. Equation (6.19) thus expresses $\mathbf{z}_t$ as a linear combination of $n-m$ random walks, plus some non-integrated 'transitory' components, $\mathbf{a}_t$, and may be regarded as a multivariate extension of the Beveridge and Nelson (1981) decomposition introduced in chapter 3.2.1.

Stock and Watson (1988b) refer to this as the *Common Trends* represen-

Table 6.1 *Likelihood ratio test statistics for common stochastic trends*

		Quantiles	
m	$-2\ln Q_m$	90%	95%
3	0.12	2.9	4.2
2	2.52	10.3	12.0
1	10.58	21.2	23.8
0	22.06	35.6	38.6

tation of $\mathbf{z}_t$, since the n-dimensional vector $\mathbf{z}_t$ can be represented by just $n-m$ 'common' stochastic trends, and they use it to develop tests of a hypothesised number of common trends and hence, equivalently, of a hypothesised number of cointegrating vectors.

Example 6.1: Are bond markets cointegrated?

In example 5.6 we analysed a VAR containing the first differences of four bond market series. Prior analysis in Mills and Mills (1991) has established that each series is $I(1)$, so that multivariate modelling of their differences is appropriate as long as there does not exist a cointegrating relationship between the variables.

On estimating the cointegrating regression between the four series in levels (normalising on the UK), the *dw* statistic was found to be 0.03 while the $\hat{\tau}_\mu$ statistic was computed to be -2.70, neither of which is large enough to reject the null hypothesis of non-cointegration in favour of the alternative of cointegration. This is consistent with the results obtained from using Johansen's (1988) LR test, calculated assuming a lag length of $p=8$, which are shown in table 6.1. There may be a maximum of $m=3$ cointegrating vectors, but we cannot reject the hypothesis that $m=0$, i.e., that there are no common stochastic trends and the system contains four unit roots.

The cointegrating regression is thus spurious in the sense of Phillips (1986): the regression residual is an $I(1)$ process and there is no equilibrium in the levels of the series. It is interesting to note, moreover, that the R^2 from the regression is 0.70: a value that is as predicted by the analytical results of section 6.2.

Example 6.2: Cointegration between equity prices, dividends and gilt yields in the UK

The relationship between equity prices, dividends and gilt edged stocks was once felt by market practitioners in the UK, primarily in the

1950s, to be of primary importance for forecasting future movements in prices. Analysis of this relationship dropped out of fashion in the mid 1960s, both because equities began to be regarded as a hedge against inflation, rather than as a risky alternative to gilts, and also because of the promulgation of the random walk hypothesis. Interest in the relationships existing between indices of equity prices, dividends and gilts has recently been rekindled, and Mills (1991b), amongst other things, considers whether the indices are cointegrated.

Figure 6.1 plots the logarithms of end-month observations from January 1969 to May 1989 on the *FT-Actuaries 500* equity index, the associated dividend index, and the Par Yield on twenty-year British Government Stocks: the *500* index is a major constituent of the *All Share* index analysed in various other examples, and the respective price and dividend series thus closely covary. The three series, denoted p_t, d_t and r_t respectively, are established to be $I(1)$ by a sequence of unit root tests reported in Mills (1991b). Employing, in this context, a natural normalisation, the cointegrating regression is

$$p_t = 1.05 + 1.01d_t - 1.11r_t + \hat{u}_t.$$

The *dw* statistic from this regression is $dw = 0.35$ and, with a sample of $T = 245$, this is sufficiently large to reject the null hypothesis of no cointegration. This is confirmed by the $\hat{\tau}_\mu$ statistic computed from the cointegrating residuals, which yields the value $\hat{\tau} = -4.85$, which is significant at the 0.01 level.

We thus conclude that the three series are cointegrated, but it is possible that there are $m = 2$ cointegrating vectors. LR statistics for testing the sequence of null hypotheses $m \leq 2$, $m \leq 1$ and $m = 0$ are 2.9, 7.6 and 32.2, which should be compared with the 0.05 critical values 4.2, 12.0 and 23.8 respectively, thus confirming that there is just one cointegrating vector.

6.5 Alternative representations of cointegrated systems

In the above discussion of testing for cointegration, various equivalent representations of a cointegrated system have been introduced, viz. the VAR representation of the levels $\mathbf{z}_t$, (6.14), and associated *interim multiplier* representation (6.15), the moving average representation for the differences $\nabla \mathbf{z}_t$, (6.16), and the common trends representation (6.19). Yet another representation is the autoregressive *error correction model* (ECM) formulation. This is obtained by rewriting the VAR(p) representation (6.14)

$$\mathbf{A}(B)\mathbf{z}_t = \mathbf{e}_t,$$

where $\mathbf{A}(B) = \mathbf{I} - \mathbf{A}_1 B - \ldots - \mathbf{A}_p B^p$, as

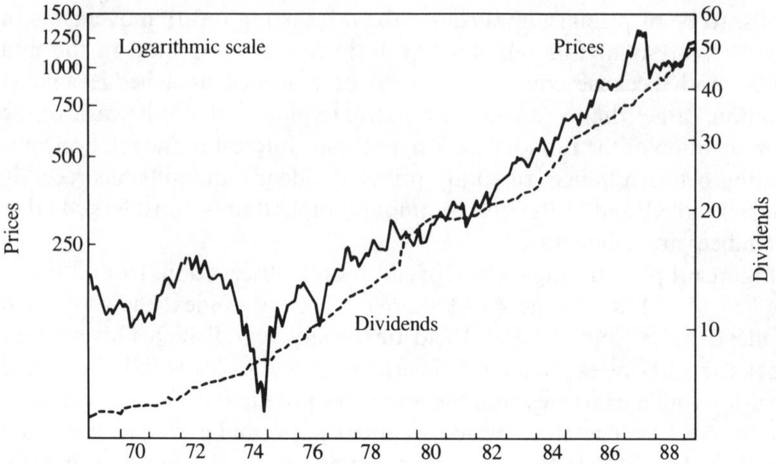

Figure 6.1A FTA 500 index: prices and dividends (monthly January 1969–May 1989)

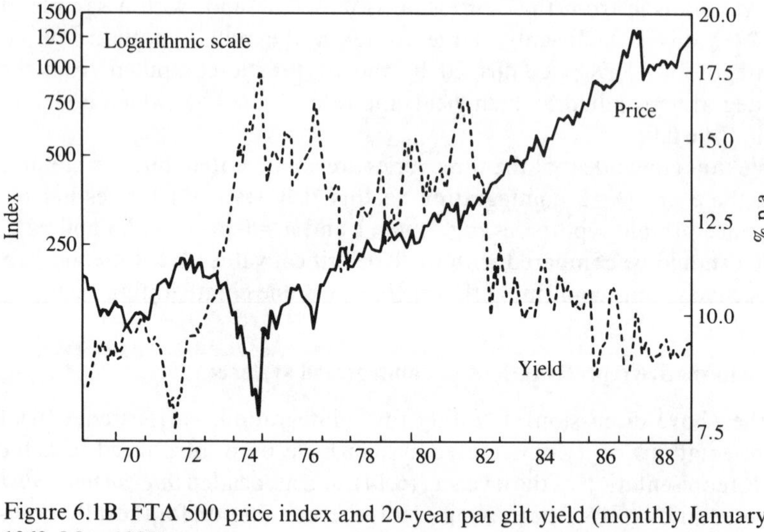

Figure 6.1B FTA 500 price index and 20-year par gilt yield (monthly January 1969–May 1989)

$$((1-B)\mathbf{A}^*(B)+\mathbf{A}(1))\mathbf{z}_t=\mathbf{e}_t,$$

where $\mathbf{A}^*(B)=\mathbf{I}-\mathbf{A}_1^*B-\ldots-\mathbf{A}_{p-1}^*B^{p-1}$, $\mathbf{A}_i^*=\sum_{i+1}^{p}\mathbf{A}_j$, $(i=1,2,\ldots,p-1)$. This can be written equivalently as

$$(1-B)\mathbf{A}^*(B)\mathbf{z}_t=-\mathbf{A}(1)\mathbf{z}_{t-1}+\mathbf{e}_t=-\gamma\alpha'\mathbf{z}_{t-1}+\mathbf{e}_t,$$

i.e., as

$$\nabla\mathbf{z}_t=\sum_{i=1}^{p-1}\mathbf{A}_i^*\nabla\mathbf{z}_{t-i}+\gamma\mathbf{u}_{t-1}+\mathbf{e}_t,$$

where $\mathbf{u}_t=\alpha'\mathbf{z}_t$ is the 'equilibrium error'. This representation may be regarded as a multivariate formulation of the error correction models popularly used in dynamic econometric modelling in which a proportion, γ, of the disequilibrium, $\mathbf{u}_{t-1}$, in one period is corrected in the next period: see Davidson *et al.* (1978) and Hendry (1987).

The ECM formulation is obtained by assuming that the innovation $\mathbf{u}_t$ in (6.6) has a moving average representation which can be approximated by a pth order VAR. Phillips (1991c) and Phillips and Loretan (1991) consider an alternative formulation of a cointegrated system having the generating process

$$y_t=\beta'\mathbf{x}_t+u_{1t}, \tag{6.20}$$

$$\nabla\mathbf{x}_t=\mathbf{u}_{2t}, \tag{6.21}$$

where we use the partitions $\mathbf{z}_t'=(y_t',\mathbf{x}_t')$ and $\mathbf{u}_t'=(u_{1t}',\mathbf{u}_{2t}')$. This system is in structural equation format with (6.20) being regarded as a stochastic version of the partial equilibrium relationship $y_t=\beta'\mathbf{x}_t$ with u_{1t} representing stationary deviations from equilibrium. Equation (6.21) is a reduced form specifying $\mathbf{x}_t$ as a general integrated process.

Taking differences of (6.20) we obtain

$$\nabla y_t=\beta'\mathbf{u}_{2t}+\nabla u_{1t}=-u_{1t-1}+(1,\beta')\mathbf{u}_t,$$

which, when combined with (6.21), yields the system

$$\nabla\mathbf{z}_t=-e\alpha'\mathbf{z}_{t-1}+\mathbf{v}_t, \tag{6.22}$$

where

$$e'=(1,\mathbf{0}),\quad \alpha'=(1,\beta'),\quad \mathbf{v}_t=\begin{bmatrix}1 & \beta'\\ 0 & \mathbf{I}\end{bmatrix}\mathbf{u}_t.$$

Equation (6.22) is a *triangular system* ECM and explains the differences $\nabla\mathbf{z}_t$ in terms of the lagged levels $\mathbf{z}_{t-1}$ and the stationary errors $\mathbf{v}_t$, these being a simple transform of the original errors $\mathbf{u}_t$.

Phillips and Loretan (1991) contrast and compare the autoregressive and triangular ECM systems (6.19) and (6.22), preferring the latter because (i) the model is linear rather than non-linear in α; (ii) the vector e is specified and not estimated: this achieves identification of the system and, because of this, the first coefficient of e cannot be interpreted as an adjustment parameter because it is no longer natural to think of (6.20) as explaining the extent of the adjustment towards equilibrium each period. Rather, the equation is to be thought of as explaining the *stationary* deviations about equilibrium that persist from period to period; and (iii) all the transient (short-run) dynamics are absorbed into the residual process $\mathbf{v}_t$ and the resulting triangular structure simplifies issues of inference.

In the autoregressive ECM, α appears non-linearly and the coefficient vector γ is unknown and must be estimated, so that (6.19) is effectively a non-linear-in-parameters reduced form. Identification is achieved by the reduction of the error process to a martingale difference sequence through assuming an autoregressive operator $\mathbf{A}(B)$: theoretically this is of infinite order and must therefore be approximated using, for example, order-selection criteria. Note that these autoregressive coefficients must be estimated simultaneously with γ and α. The autoregressive ECM does allow, though, testing of the row rank of $\mathbf{B}$ and hence estimation of the number of cointegrating vectors.

Neither of these specifications corresponds to the single-equation empirical specification of ECMs that follow in the Davidson *et al.* (1978) tradition. Such specifications concentrate attention on parametric representations of (6.20) alone, ignoring the generating process (6.21), and either augment the equilibrium relationship $y_t = \beta'\mathbf{x}_t$ with stationary variables designed to capture the short-run disequilibrium dynamics, viz.

$$y_t = \beta'\mathbf{x}_t + \sum_{i=1}^{p} f_{1i}\nabla y_{t-i} + \sum_{i=0}^{p} \mathbf{f}'_{2i}\nabla \mathbf{x}_{t-i} + e_t, \tag{6.23}$$

or consider a single equation version of the autoregressive ECM,

$$\nabla y_t = f_0(y_{t-1} - \beta'\mathbf{x}_{t-1}) + \sum_{i=1}^{p} f_{1i}\nabla y_{t-i} + \sum_{i=0}^{p} \mathbf{f}'_{2i}\nabla \mathbf{x}_{t-i} + e_t. \tag{6.24}$$

Formulations of this type are referred to by Phillips and Loretan (1991) as *Single-equation ECMs* (SEECMs).

6.6 Estimation of cointegrated systems

Estimation of, and inference in, the various representations of cointegrated systems has been studied extensively by Phillips (1988, 1991c), Phillips and Hansen (1990) and Phillips and Loretan (1991). The simplest approach is to estimate (6.20) by LS: this has already been considered when using such a method to estimate the cointegrating regression prior to testing the resultant residuals for cointegration. Although $\hat{\beta}$ is $O(T^{-1})$ consistent, in general, i.e., when u_{1t} and $\mathbf{u}_{2t}$ are not independent, so that $\mathbf{x}_t$ is endogenous, it will be asymptotically biased and the limit distribution will be asymmetric and contain nuisance parameters. These defects are only partially remedied by using single-equation structural methods like two-stage LS, and so all such methods may be regarded as poor candidates for inference.

These deficiencies have led Phillips and Hansen (1990) to propose semiparametric corrections for serial correlation and endogeneity to the LS estimator $\hat{\beta}=(\mathbf{X}'\mathbf{X})^{-1}\mathbf{X}'\mathbf{y}$, where

$$\mathbf{X}=(\mathbf{x}_1,\ldots,\mathbf{x}_T) \text{ and } \mathbf{y}=(y_1,\ldots,y_T).$$

These modifications use consistent estimates of the covariance matrices Σ and $\Delta=\Sigma_0+\Sigma_1$, obtained from the cointegrating regression of y_t on $\mathbf{x}_t$ (recall equation (6.7)), to construct

$$y_t^+=y_t-\hat{\sigma}_{21}'\hat{\Sigma}_{22}^{-1}\nabla\mathbf{x}_t$$

and

$$\hat{\delta}^+=\hat{\Delta}\begin{bmatrix}1\\-\hat{\Sigma}_{22}^{-1}\hat{\sigma}_{21}\end{bmatrix},$$

where $\hat{\Sigma}_{22}$ and $\hat{\sigma}_{21}$ are appropriate submatrices of $\hat{\Sigma}$ and $\hat{\Delta}$. Using these, the *fully modified* LS estimator is given by

$$\hat{\beta}^+=(\mathbf{X}'\mathbf{X})^{-1}(\mathbf{X}'\mathbf{y}^+-T\hat{\delta}^+).$$

Phillips and Hansen (1990) show that, once an appropriate estimate of the variance matrix of $\hat{\beta}^+$ is constructed, conventional asymptotically normal inference may be carried out. The fully modified LS estimator has the same asymptotic behaviour as the full systems ML estimator $\tilde{\beta}$, which is obtained by applying LS to the augmented equation (Phillips and Loretan, 1991, Saikkonen, 1991)

$$y_t=\beta'\mathbf{x}_t+\sum_{i=-K_1}^{K_2}\phi_i\nabla\mathbf{x}_{t-i}+u_{1.2t}, \tag{6.25}$$

where the number of leads (K_1) and lags (K_2) are chosen so as to reduce the regression error $u_{1.2t}$ to approximately white noise. From Phillips (1991c), $\tilde{\beta}$ is consistent and asymptotically median unbiased and its limit distribution involves only scale parameters, which are readily eliminated by the use of conventional test statistics: for example, asymptotic χ^2 criteria are obtained from the usual construction of Wald, LM and LR tests.

Saikkonen (1991) also shows that efficiency is improved if leads and lags of the differences of other $I(1)$ variables, which do not appear in the cointegrating regression, are included in (6.25). An example of this is the extension considered by Phillips and Loretan (1991), in which lagged values of the equilibrium error $y_t - \beta'\mathbf{x}_t$ are included

$$y_t = \beta'\mathbf{x}_t + \sum_{i=-K_1}^{K_2} \phi_i \nabla \mathbf{x}_{t-i} + \sum_{i=1}^{K_3} \theta_i (y_{t-i} - \beta'\mathbf{x}_{t-i}) + v_t, \tag{6.26}$$

this choice being shown to follow from assuming that the generating process for $\mathbf{u}_t$ is a vector autoregression. Note that β appears non-linearly and it is thus non-linear LS (NLS) that provides an asymptotically consistent and efficient estimate of long-run equilibrium, $\tilde{\beta}^*$ say.

Most attention has been concentrated on the estimation of SEECMs. Comparison of equations (6.23) and (6.26) shows that they differ in that the former uses lagged values of ∇y_t in place of the lead values of $\nabla \mathbf{x}_t$ and lagged values of $y_t - \beta'\mathbf{x}_t$ contained in the latter. Phillips and Loretan (1991) argue that, if there is feedback from u_1 to $\mathbf{u}_2$, *leads* of $\nabla \mathbf{x}_t$ must be included in SEECM specifications to obtain errors that form a martingale difference sequence with respect to the past history of u_1 and the full history of $\mathbf{u}_2$. There is also an asymptotic advantage to the use of lagged disequilibrium terms rather than lags of ∇y_t, as the latter are not, in general, an adequate proxy for the past history of u_1 because of the persistence in the effects of the innovations that arises from the presence of unit roots in the system. This is most easily seen by supposing that $K_1 = -1$ in (6.26), so that there is no feedback and hence $\mathbf{x}_t$ is strongly exogenous. By setting initial conditions $y_0 = \mathbf{x}_0 = 0$, we may write

$$y_{t-1} = \sum_{s=1}^{t-1} \nabla y_{t-s}, \qquad \mathbf{x}_{t-1} = \sum_{s=1}^{t-1} \nabla \mathbf{x}_{t-s},$$

and substituting these into (6.26) then obtains the SEECM (6.23) but with the lag length p set equal to $t-1$. Moreover, the lag coefficients f_{1i} and $\mathbf{f}_{2i}$ do not, in general, decay as the lag increases because the above partial sums imply unit weights for individual innovations. Thus, in order to model short-run dynamics using the variables ∇y_{t-i} and $\nabla \mathbf{x}_{t-i}$, it is necessary to include all lags because of this shock persistence, which is quite impractical in applications and cannot be justified in theory where truncation arguments are needed to develop the asymptotics.

Asymptotic theory thus favours the use of NLS on non-linear-in-parameters SEECM's rather than LS on linear SEECM models formulated with lags (and possibly leads) of differences in all variables in the system. Simulation evidence presented in Phillips and Loretan (1991) provides considerable support for this argument in terms of the small-sample performance of the procedures. Indeed, the NLS SEECM estimator tends to outperform the fully modified LS estimator, both of which are clearly preferable to 'unmodified' LS.

Because NLS has computational disadvantages, Phillips and Loretan (1991) suggest that a two-stage modelling procedure may be preferable. This is to (i) estimate efficiently the long-run equilibrium relationships by system methods (either fully modified LS or ML), and then to (ii) utilise the estimates from stage (i) in the construction of 'parsimoniously parameterised' ECM's. This has the additional advantage that it would not be necessary to include leads in the ECM regression since the long-run equilibrium has already been efficiently estimated and incorporated.

Example 6.3: Modelling the relationship between equity prices, dividends and gilt yields as a cointegrated system

Example 6.2 has established that equity prices, dividends and gilt yields in the UK are cointegrated and thus enables the relationship between the three series to be modelled as a cointegrated system. From the cointegrating regression reported in that example, the LS estimator of the cointegrating vector is $\hat{\beta}'=(1.01, -1.11)$. Although consistent, $\hat{\beta}$ is asymptotically biased and non-normal when d_t and r_t are endogenous and u_t is serially correlated.

ML estimates were obtained by incorporating various lags and leads of ∇d_t and ∇r_t. With just current values of these variables introduced, i.e., setting $K_1=K_2=0$ in (6.25), we obtained identical estimates to those of $\hat{\beta}$ reported above, and examining a range of K_1 and K_2 settings produced no discernible alteration to these estimates, thus suggesting that any endogeneity of d_t and r_t has little effect here on the estimation of the cointegrating vector. Further modelling was thus restricted to using just lags of ∇d_t and ∇r_t.

Attention was then focused on building a SEECM. Initially, a model of the form (6.23) was attempted. However, this was unsuccessful for the reasons discussed above in section 6.6: the lag coefficients failed to die out, remaining significant at high lags, and the error term was not reduced to white noise – exactly the problems encountered by the shock persistence caused by the unit roots in the system.

A non-linear model of the form (6.26) was then constructed, with a parsimonious parameterisation being obtained in the usual 'à la Hendry' fashion by eliminating regressors because of insignificant coefficients, etc. This produced the following model

$$\begin{aligned} p_t = {} & \underset{(0.50)}{0.49} + \underset{(0.05)}{1.07}d_t - \underset{(0.14)}{1.03}r_t + \underset{(0.03)}{0.85}(p_{t-1} - 0.49 - 1.07d_{t-1} + 1.03r_{t-1}) \\ & - \underset{(0.35)}{1.12}\nabla d_t - \underset{(0.33)}{0.84}\nabla d_{t-3} + \underset{(0.15)}{0.44}\nabla r_t - \underset{(0.09)}{0.14}\nabla r_{t-1}, \end{aligned}$$

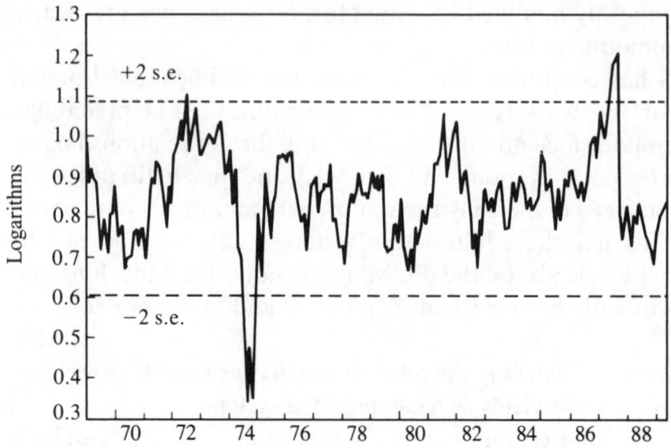

Figure 6.2 'Confidence' (monthly January 1969–May 1989)

where asymptotic standard errors are shown in parentheses. The long-run equilibrium is thus augmented by the last periods equilibrium error and by current and past changes in d_t and r_t, these variables thus modelling the stationary deviations around equilibrium, i.e., the short-run dynamics of the system. Note that this equation can be rewritten as

$$\nabla p_t = -0.15(p_{t-1} - 0.49 - 1.07d_{t-1} + 1.03r_{t-1})$$
$$-0.05\nabla d_t - 0.84\nabla d_{t-3} + 0.59\nabla r_t - 0.14\nabla r_{t-1},$$

which is in the more familar ECM form. It is interesting to note that this form of the model is almost identical, apart from the deletion of the statistically insignificant ∇d_t and the inclusion of an additional ∇p_{t-3} term, with the equation for ∇p_t reported in Mills (1991b), which was obtained through a two-stage modelling of the type suggested by Phillips and Loretan (1991).

The analysis of Mills (1991b) was set within a VAR framework, leading to an autoregressive ECM model for the three variables. The resulting model was then investigated in terms of its dynamic multipliers and impulse response functions (see chapter 5.6): Mills (1991b) provides details.

The estimate of the cointegrating vector is very close to $\beta' = (1, -1)$ and tests of this null hypothesis prove not to be significant: for example, a LR statistic calculated from the non-linear SEECM above produced the value of 1.15, asymptotically distributed as χ^2_2. Under such a restriction, the long-run equilibrium becomes $p_t = \alpha + d_t - r_t$ or, in terms of the levels, P_t, D_t and R_t, of the series,

$$C_t = R_t/(D_t/P_t),$$

where $C_t = \exp(\alpha)$. Since D_t/P_t is the dividend yield, in equilibrium the gilt yield and dividend yield are in constant proportion to each other. Moreover, deviations from this equilibrium are stationary, so that divergences from the ratio can only be temporary.

This ratio is, in fact, exactly the decomposition used by investment analysts of the 1950s and early 1960s to analyse movements in equity prices and was termed by them the *confidence factor*. Figure 6.2 presents a plot of the logarithm of C_t for the period from 1969 to 1989. Mills (1991b) investigates the series in more detail and shows that it is generated by an AR(1)–ARCH(5) process (see chapter 4.4). From this model an unconditional variance can be calculated, and the resulting two-standard error bounds are superimposed on the plot. Although standard probability statements should not be attached to these bounds because the presence of ARCH renders the distribution non-Gaussian, having fatter tails than the normal, they are, nevertheless, seen to act as 'resistance lines': only during the latter months of 1974 and 1987, both periods of great upheaval in the UK equity market, are they broken more than momentarily, at almost all other times their penetration is immediately reversed.

7 Further topics in the multivariate modelling of financial time series

While many of the topics introduced in previous chapters have applications in other areas of economics, most notably macroeconomics, certain multivariate techniques have been developed almost exclusively for use in financial applications. This chapter reviews such techniques, which are all extensions of the material developed in chapters 5 and 6. We begin by discussing multivariate GARCH-M models, which provide a methodology for estimating and testing the time-varying CAPM. We then go on, in section 2, to consider volatility tests as a means of investigating the implications of present value models: the 'first generation' of such tests use variance bounds techniques, while the later tests employ cointegration restrictions within a VAR framework peculiar to the present value setup. One of the assumptions of the present value model is that a transversality condition is assumed to hold: this rules out the possibility of 'rational bubbles'. Section 3 investigates the implications of relaxing this condition and discusses the testing procedures for rational bubbles that have been proposed recently. Finally, section 4 introduces a log-linear approximation to the present value model, the 'dividend-ratio' model, that is claimed to possess certain advantages over the more conventional formulations analysed earlier in the chapter.

7.1 Multivariate GARCH-M models

In chapter 5.2 we considered the GARCH-M regression model, a simple form of which can be written, using the notation of that chapter, as

$$\begin{aligned} y_t &= \mathbf{X}_t\beta + \delta h_t + \epsilon_t \\ h_t &= \gamma_0 + \sum_{i=1}^{p} \gamma_i \epsilon_{t-i}^2 + \sum_{i=1}^{q} \phi_i h_{t-i}, \end{aligned} \tag{7.1}$$

where $\epsilon_t \mid \Phi_{t-1} \sim N(0,h_t)$, so that

$$h_t = E(\epsilon_t^2 \mid \Phi_{t-1}) = E((y_t - \mu_t)^2 \mid \Phi_{t-1}),$$

with

$$\mu_t = E(y_t \mid \Phi_{t-1}) = \mathbf{X}_t\beta + \delta h_t .$$

$\mathbf{X}_t$ is a $k \times 1$ vector of weakly exogenous and lagged dependent variables. Engle and Bollerslev (1986) consider a multivariate extension of (7.1) for modelling the $n \times 1$ vector $\mathbf{y}_t$, (cf. chapter 5.4)

$$\mathbf{y}_t = \mathbf{B}\mathbf{X}_t + \mathbf{D}vech(\mathbf{H}_t) + \mathbf{u}_t , \tag{7.2a}$$

$$vech(\mathbf{H}_t) = \mathbf{C} + \sum_{i=1}^{p} \mathbf{A}_i vech(\mathbf{u}_{t-i}\mathbf{u}'_{t-i}) + \sum_{i=1}^{q} \mathbf{B}_i vech(\mathbf{H}_{t-i}), \tag{7.2b}$$

where $\mathbf{u}_t \mid \Phi_{t-1} \sim N(\mathbf{0},\mathbf{H}_t)$. In this formulation, $\mathbf{H}_t = E(\mathbf{u}_t\mathbf{u}'_t \mid \Phi_{t-1})$ is the $n \times n$ conditional covariance matrix associated with the error vector $\mathbf{u}_t$ and vech ($\mathbf{H}_t$) denotes the $\frac{1}{2}n(n+1) \times 1$ vector of all the unique elements of $\mathbf{H}_t$ obtained by stacking the lower triangle of $\mathbf{H}_t$. $\mathbf{B}$ and $\mathbf{D}$ are thus $n \times k$ and $n \times \frac{1}{2}n(n+1)$ matrices, $\mathbf{C}$ is an $\frac{1}{2}n(n+1) \times 1$ vector and $\mathbf{A}_1,\mathbf{A}_2,\ldots,\mathbf{A}_p$, $\mathbf{B}_1,\mathbf{B}_2,\ldots,\mathbf{B}_q$ are $\frac{1}{2}n(n+1) \times \frac{1}{2}n(n+1)$ matrices of coefficients.

Estimation of (7.2) follows standard procedures: for example, ML estimates can be obtained by maximising the likelihood function using the BHHH algorithm (for details, see Bollerslev, Engle and Wooldridge, 1988). Nevertheless, as it stands, (7.2) is very general, requiring a large number of parameters to be estimated, and various simplifications have thus been suggested. A natural simplification is to assume that each covariance depends only on its own past values and innovations, i.e., that the $\mathbf{A}_i$ and $\mathbf{B}_i$ matrices are diagonal. In addition, most applications of the model (for example, Bollerslev, Engle and Wooldridge, 1988, Hall, Miles and Taylor, 1989, and Baillie and Myers, 1991) have restricted attention to multivariate GARCH(1,1) systems, i.e., they set $p=q=1$, so that each row of the conditional variance equation (7.2b) can be written as

$$h_{ijt} = c_{ij} + a_{ij}u_{it-1}u_{jt-1} + b_{ij}h_{ijt-1}, \quad i,j = 1,2,\ldots,n .$$

An alternative to the diagonal vech parameterisation is the positive definite parameterisation, where

$$\mathbf{H}_t = \mathbf{C}'\mathbf{C} + \mathbf{A}'\mathbf{u}_{t-1}\mathbf{u}'_{t-1}\mathbf{A} + \mathbf{B}'\mathbf{H}_{t-1}\mathbf{B},$$

and this has been used by both Hall, Miles and Taylor (1989) and Baillie and Myers (1991).

Example 7.1: A CAPM with time-varying covariances

As discussed in example 5.4, the CAPM states that the expected excess return on an asset (or small portfolio), r_p, over the return on a risk-free asset, r_f, is proportional to the expected excess return on the market portfolio, r_m, the factor of proportionality being equal to the asset beta. If these expectations at time t are conditioned on the information set available to agents at time $t-1$, Φ_{t-1}, then the CAPM can be written as

$$E(r_{pt}|\Phi_{t-1})-r_{ft-1}=\beta_{pt}[E(r_{mt}|\Phi_{t-1})-r_{ft-1}]$$
$$\beta_{pt}=Cov(r_{pt},r_{mt}|\Phi_{t-1})/V(r_{mt}|\Phi_{t-1}),$$

where, because we now allow the covariance matrix of returns

$$\mathbf{H}_t=\begin{bmatrix} V(r_{pt}|\Phi_{t-1}) & Cov(r_{pt},r_{mt}|\Phi_{t-1}) \\ Cov(r_{pt},r_{mt}|\Phi_{t-1}) & V(r_{mt}|\Phi_{t-1}) \end{bmatrix}$$

to vary over time, both the expected returns and the betas will, in general, be time varying.

This formulation of the CAPM is, however, non-operational because of the lack of an observed series for the expected market excess return. Bollerslev, Engle and Wooldridge (1988) and Hall, Miles and Taylor (1989), henceforth referred to as BEW and HMT respectively, thus assume that the 'market price of risk', λ, defined as

$$\lambda=[E(r_{mt}|\Phi_{t-1})-r_{ft-1}]/V(r_{mt}|\Phi_{t-1}),$$

is constant, so that

$$E(r_{mt}|\Phi_{t-1})-r_{ft-1}=\lambda\cdot V(r_{mt}|\Phi_{t-1}).$$

Hence, we can write

$$r_{pt}=r_{ft-1}+\lambda\cdot Cov(r_{pt},r_{mt}|\Phi_{t-1})+u_{pt}$$
$$r_{mt}=r_{ft-1}+\lambda\cdot V(r_{mt}|\Phi_{t-1})+u_{mt},$$

where

$$u_{pt}=r_{pt}-E(r_{pt}|\Phi_{t-1})$$

and

$$u_{mt}=r_{mt}-E(r_{mt}|\Phi_{t-1}).$$

We thus see that

$$V(r_{mt}|\Phi_{t-1})=E(u_{mt}^2|\Phi_{t-1})=h_{mmt}$$

and

$$Cov(r_{pt},r_{mt}|\Phi_{t-1})=E(u_{pt}u_{mt}|\Phi_{t-1})=h_{pmt},$$

so that this time-varying CAPM can be put into multivariate GARCH-M form as

$$\mathbf{y}_t = \mathbf{b} + \mathbf{d}\,vech(\mathbf{H}_t) + \mathbf{u}_t,$$

where

$$\mathbf{y}_t = (r_{pt} - r_{ft-1}, r_{mt} - r_{ft-1})'$$
$$vech(\mathbf{H}_t) = (h_{ppt}, h_{mmt}, h_{pmt})'$$
$$\mathbf{u}_t = (u_{pt}, u_{mt})'$$

and

$$\mathbf{d} = \lambda\begin{bmatrix} 0 & 0 & 1 \\ 0 & 1 & 0 \end{bmatrix}.$$

b is a vector of constants: a non-zero vector might reflect a preferred habitat phenomenon or differential tax treatment of assets. The model is easily generalised to contain more than one asset or portfolio and models of this type have been estimated by BEW and HMT, both employing a GARCH(1,1) conditional variance structure with a diagonal vech parameterisation.

BEW use a market portfolio for the US comprising yields on six-month Treasury bills and twenty-year Treasury bonds and New York Stock Exchange equity returns, using quarterly observations from 1959 to 1984 and with the three-month Treasury bill yield as the risk-free return. They estimate the market price of risk, λ, to be almost 0.5 and find that the intercepts for bonds and stocks are significantly negative, explaining this by the fact that reduced capital gains taxes on long-term assets provide incentives to hold these assets even at otherwise unfavourable rates of return. The estimates of the conditional variance parameters are significant for bills and bonds and, although individually insignificant for stocks, are highly significant as a group. The estimates are used to calculate the risk premia $\mathbf{b} + \mathbf{d}\,vech(\mathbf{H}_t)$ and the implied time-varying betas. Bills and bonds have rising risk premia after October 1979, the beta for stocks is close to one, that for bonds slightly above one, and that for bills close to zero, although there is substantial movement in the latter two series over the sample period.

HMT construct four portfolios from monthly data on the London Stock Exchange over the period 1975 to 1987, using returns on the *FTA 500* index as the market return and the Treasury Bill yield as the risk-free return. Their preferred model yields an estimate of λ of 3.25, and the interesting finding that the matrix **C** in the GARCH equation is diagonal: they interpret this as implying that agents expect a non-zero variance on all portfolios but have no prior expectation about the long-run covariances of the system. They impose in estimation the restriction that $\mathbf{A}_1$ and $\mathbf{B}_1$ are scalars, so that agents are assumed to attach the same relative importance to past events in forming expectations about price volatility in each portfolio. The estimates of these parameters are highly significant, with $\hat{\mathbf{A}}_1+\hat{\mathbf{B}}_1$ being 0.986, implying a mean lag of around twenty-two months. Again time-varying betas are calculated and reported.

HMT extend the model along the lines of Breedon's (1979) consumption-based CAPM by replacing the expected return equation with

$$r_{pt}=r_{ft-1}+\lambda\cdot Cov(r_{pt},r_{mt}\mid\Phi_{t-1})+\lambda_c\cdot Cov(r_{pt},\nabla c_t\mid\Phi_{t-1})+u_{pt},$$

where ∇c_t is the change in the logarithm of real consumption and λ_c is the price of consumption-related risk. Consequently, they include an additional equation for ∇c_t, a random walk with drift specification, with the conditional variance equation also being suitably modified. Both risk parameters are found to be significantly positive, implying that the original CAPM and its consumption-based variant are outperformed by a more general model incorporating both measures of risk. This extension significantly alters the structure of the GARCH process, however. The weight attached to past history is now much smaller, with $\hat{\mathbf{A}}_1+\hat{\mathbf{B}}_1$ being less than 0.5, although both are still significantly positive so that time variation is still present in conditional variances and covariances and hence risk premia. **C** is estimated to contain a number of significant off-diagonal elements, thus implying that unconditional covariances are now non-zero.

7.2 Present value models, excess volatility and cointegration

7.2.1 Present value models and the 'simple' efficient markets hypothesis

As remarked in chapter 1, present value models are extremely popular in finance as they are often used to formulate models of efficient markets. Written generally, a present value model for two variables, y_t and x_t, states that y_t is a linear function of the present discounted value of the expected future values of x_t

$$y_t=\theta(1-\delta)\sum_{i=0}^{\infty}\delta^{i+1}E(x_{t+i}\mid\Phi_t)+c, \tag{7.3}$$

where c, the constant, θ, the coefficient of proportionality, and δ, the constant discount factor, are parameters that may be known a priori or may need to be estimated. As usual, $E(x_{t+i}\mid\Phi_t)$ is the expectation of x_{t+i} conditional on the information set available at time t, Φ_t.

A simple example of how (7.3) might arise is to consider an implication of the efficient markets hypothesis: that stock returns, r_t, are unforecastable, i.e., $E(r_{t+1}\mid\Phi_t)=r$, where r is a constant, sometimes referred to as the discount rate (see Shiller 1981a, 1981b). If y_t is the beginning of period t stock price and x_t the dividend paid during the period, then

$$r_{t+1}=(y_{t+1}-y_t+x_t)/y_t,$$

so that we can express y_t as a first-order rational expectations model of the form

$$y_t=\delta E(y_{t+1}\,|\,\Phi_t)+\delta E(x_t\,|\,\Phi_t), \tag{7.4}$$

where $\delta=1/(1+r)$. This can be solved by recursive substitution to yield

$$y_t=\sum_{i=0}^{n}\delta^{i+1}E(x_{t+i}\,|\,\Phi_t)+\delta^n E(y_{t+n}\,|\,\Phi_t). \tag{7.5}$$

If we impose the terminal (or transversality) condition that the second term in (7.5) goes to zero as $n\rightarrow\infty$, the present value relation (7.3) is obtained with $c=0$ and $\theta=1/(1-\delta)$. An additional implication of the model is obtained by noting that $E(x_t\,|\,\Phi_t)=x_t$, so that (7.4) can be written as

$$y_t=\delta x_t+\delta E(y_{t+1}\,|\,\Phi_t).$$

Defining the innovation in y_t to be the change in the conditional expectation of y_t that is made in response to new information arriving between $t-1$ and t, i.e.

$$\tilde{\Delta}_t y_t=E(y_t\,|\,\Phi_t)-E(y_t\,|\,\Phi_{t-1}),$$

then, since $E(y_t\,|\,\Phi_{t-1})=y_{t-1}/\delta-x_{t-1}$ and $E(y_t\,|\,\Phi_t)=y_t$, this innovation is

$$\tilde{\Delta}_t y_t=y_t-y_{t-1}/\delta+x_{t-1}=\nabla y_{t-1}+(x_{t-1}-ry_{t-1}),$$

so that, in the context of the stock price and dividend example, it is the change in price corrected for price movements in response to short-run dividend variations. Now, by the rule of iterated expectations

$$\begin{aligned}E(\tilde{\Delta}_t y_{t+k}\,|\,\Phi_{t-m})&=E(E(y_{t+k}\,|\,\Phi_t)-E(y_{t+k}\,|\,\Phi_{t-1})\,|\,\Phi_{t-m})\\&=E(y_{t+k}\,|\,\Phi_{t-m})-E(y_{t+k}\,|\,\Phi_{t-m})=0\end{aligned}$$

for $m>0$, so that $\tilde{\Delta}_t y_{t+k}$ must be uncorrelated for all k with information known at time $t-1$ and must, since lagged innovations are information at time t, be uncorrelated with $\tilde{\Delta}_s y_{t+j}$, $s<t$, all j, i.e., innovations are serially uncorrelated.

Shiller (1981a) also emphasises an alternative form of the model that is based on the *perfect foresight* or *ex post rational* series y_t^*, defined as the present value of *actual* subsequent values of x_t

$$y_t^*=\theta(1-\delta)\sum_{i=0}^{\infty}\delta^{i+1}x_{t+i}+c. \tag{7.6}$$

The present value model thus states that actually observed y_t must be the conditional expectation of y_t^*, i.e., $y_t = E(y_t^* \mid \Phi_t)$.

It should be emphasised that the present value model (7.3) restricts the discount rate r, and associated discount factor $\delta = 1/(1+r)$, to be constant. An alternative is to allow these parameters to be time varying, so that we have $\delta_t = 1/(1+r_t)$. Abstracting from coefficients of proportionality and constant terms, the present value model then becomes

$$y_t = E\left(\sum_{i=0}^{\infty} \left(\prod_{j=0}^{i} \delta_{t+j} \right) x_{t+i} \middle| \Phi_t \right),$$

which has the implication that *discounted* returns $\delta_{t+1} r_{t+1}$ should not be forecastable.

7.2.2 *Volatility tests*

Since the present value-efficient markets model provides a precise description of the behaviour of returns and innovations, it is not surprising that it has been subjected to much empirical testing.

The 'first-generation' of tests concentrated on the implications of the model for the relative variability of the series involved, y_t, y_t^* and x_t: hence their description as *volatility tests*. The assertion that $y_t = E(y_t^* \mid \Phi_t)$ implies that $V(y_t) + V(u_t) = V(y_t^*)$, where $u_t = y_t^* - y_t$ is the forecast error which, by virtue of y_t being an optimal forecast of y_t^*, will be uncorrelated with y_t, so that $Cov(y_t, u_t)$ will be zero. In terms of standard deviations, we have, on dropping subscripts, i.e., by implicitly assuming stationarity of y_t, y_t^* and x_t (LeRoy and Porter, 1981, Shiller, 1979, 1981a)

$$\sigma(y) \leq \sigma(y^*). \tag{7.7}$$

Although most attention has been focused on (7.7), Shiller (1981a, 1981b) derives two further inequalities. The first puts a limit on the standard deviation of ∇y in terms of the standard deviation of x:

$$\sigma(\nabla y) \leq \sigma(x)/\sqrt{2r}, \tag{7.8}$$

while the second puts a limit on the standard deviation of ∇y in terms of the standard deviation of ∇x

$$\sigma(\nabla y) \leq \sigma(\nabla x)/\sqrt{2r^3/(1+2r)}. \tag{7.9}$$

Shiller (1981b) provides intuitive interpretations of all three inequalities but, in order to apply them empirically, certain problems are encountered. The first, and perhaps the most obvious, is that y_t^* is not observable without error, since the summation in (7.6) extends to infinity. Shiller (1981a) suggests choosing an arbitrary terminal value of y_t^* based on the observed sample $\{y_t\}_1^T$, for example setting $y_T^* = y_T$ or $y_T^* = \bar{y}$: y_t^*can then be determined recursively by $y_t^* = \delta(y_{t+1}^* + x_t)$, working backwards from T. Given a computed *ex post* rational series, sample estimates of the various standard deviations can then be calculated and the inequalities examined. There are, however, a number of less transparent problems with this approach, and these form the basis of the Flavin (1983) and Kleidon (1986a, 1986b) critiques of such volatility 'tests'.

To consider the issues involved, it is instructive to analyse the model when an explicit stochastic process is assumed for x_t. Both Flavin and Kleidon assume that x_t follows a stationary AR(1) process

$$x_t = \mu_x + \phi x_{t-1} + a_t, \qquad (7.10)$$

where $|\phi| < 1$ and $a_t \sim SWN(0, \sigma_a^2)$. If the information set is assumed to comprise current and past values of x_t, Kleidon (1986a, propositions 1 and 2) shows that y_t will be a linear function of x_t

$$y_t = \frac{\mu_x \alpha}{\delta r \phi} + \alpha x_t,$$

where $\alpha = \phi/(1 + r - \phi)$, so that y_t will follow the AR(1) process

$$y_t = \mu_y + \phi y_{t-1} + b_t,$$

where $\mu_y = \mu_x/r$ and $b_t = \alpha a_t \sim SWN(0, \sigma_b^2)$, on defining $\sigma_b^2 = \alpha^2 \sigma_a^2$. y_t^* will then follow the AR(2) process

$$y_t^* = \mu^* + \phi_1 y_{t-1}^* + \phi_2 y_{t-2}^* + c_t,$$

where $\mu^* = \delta\mu_x$, $\phi_1 = \phi + \delta$, $\phi_2 = -\phi\delta$ and $c_t \sim SWN(0, \sigma_c^2)$, on defining $\sigma_c^2 = \delta^2 \sigma_a^2$. From the stationarity conditions of chapter 2.3, y_t^* will be stationary if $|\phi| < 1$ and $\delta < 1$, i.e., for $r > 0$.

The unconditional mean and variance of y_t are

$$E(y_t) = \frac{\mu_y}{1 - \phi} = \frac{\mu_x}{r(1 - \phi)},$$

and

$$V(y_t) = \sigma_b^2 \frac{1}{1 - \phi^2} = \frac{r^2 \phi^2 \sigma_a^2}{\alpha^2 (1 - r^2)},$$

while those for y_t^* are

$$E(y_t^*) = \frac{\mu^*}{(1 - \phi_1 - \phi_2)} = \frac{\delta \mu_x}{(1 - \delta)(1 - \phi)} = \frac{\mu_x}{r(1 - \phi)},$$

which equals $E(y_t)$ and, again from chapter 2.3,

$$V(y_t^*) = \left(\frac{1 - \phi_2}{1 + \phi_2}\right) \frac{\sigma_c^2}{((1 - \phi_2)^2 - \phi_1^2)} = \frac{\sigma_a^2 (1 + r + \phi)}{(1 + r - \phi)(2r + r^2)(1 - \phi^2)}.$$

The inequality (7.7) is thus readily verified since

$$\Delta = V(y_t^*) - V(y_t) = \frac{\sigma_a^2(1+r)^2}{(1+r-\phi)^2(2r+r^2)}.$$

so that $\Delta > 0$ for $r > 0$ and $\sigma_a^2 > 0$.

Although the variance inequality (7.7) is almost always given in terms of unconditional variances, Kleidon (1986b) shows that similar inequalities hold for the conditional variances:

$$V(y_t \mid \Phi_{t-k}) = V(y_t \mid y_{t-k}) = \sigma_a^2 \alpha^2 \left(\frac{1-\phi^{2k}}{1-\phi^2} \right),$$

and

$$V(y_t^* \mid \Phi_{t-k}) = V(y_t^* \mid y_{t-k}) = \sigma_a^2 \alpha^2 \left[\left(\frac{1-\phi^{2k}}{1-\phi^2} \right) + \frac{(1+r^2)}{\phi^2(2r+r^2)} \right]$$

$$= V(y_t \mid y_{t-k}) + \frac{\sigma_a^2 \alpha^2 (1+r)^2}{\phi^2(2r+r^2)},$$

i.e.

$$V(y_t^* \mid \Phi_{t-k}) \geq V(y_t \mid \Phi_{t-k}), \quad k = 0, \ldots, \infty, \tag{7.11}$$

where Φ_{t-k} is the information set at $t-k$ and is included in Φ_t. The limit as $k \to \infty$ of these conditional variances are the unconditional variances used in (7.7).

Before examining the properties of volatility tests based on the variance inequality (7.7), a related question needs to be addressed. It is often argued that one of the implications of this inequality is that the time series plot of y_t should be 'smoother' than that of y_t^*: that it very often is not in empirical applications has been perhaps the major reason why volatility tests of this type have had such a dramatic impact. Indeed, such plots often seemed so convincing that they substituted for formal statistical evidence: see, for example, Tirole (1985).

The drawing of such an implication would be unwarranted, however. If, like Kleidon (1986a, 1986b), we take the smoothness, or amount of short-term variation, in y_t and y_t^* to be determined by the variance conditional on *past values of the series*, i.e., $V(y_t \mid y_{t-k})$ and $V(y_t^* \mid y_{t-k}^*)$, then we see that the latter conditional variance is *not* equivalent to $V(y_t^* \mid \Phi_{t-k})$ since, by the definition of y_t^*, past values of y_t^* depend on *future* values of x_t, which are not known at or prior to time t. Consequently, there is no requirement for the conditional variances $V(y_t \mid y_{t-k})$ and $V(y_t^* \mid y_{t-k}^*)$ to satisfy any inequality such as (7.11), and indeed they do not. Kleidon (1986b, propositions 2 and 5) shows that, for the AR(1) case considered above

$$V(y_t \mid y_{t-k}) = \frac{\sigma_a^2 \phi^2 (1-\phi^{2k})}{(1+r-\phi)^2(1-\phi^2)}, \tag{7.12}$$

and

$$V(y_t^* \mid y_{t-k}^*) = V(y_t^*)(1-\rho_k^{*2}), \tag{7.13}$$

where ρ_k^* is the kth autocorrelation of y_t^*, given by

$$\rho_k^* = \frac{\phi^{k+1}(2r+r^2)-(1-\phi^2)(1+r^{k-1})}{(1+r+\phi)(\phi+r\phi-1)}.$$

Examination of (7.12) and (7.13) shows that the limit as $k \rightarrow \infty$ gives the unconditional variances $V(y_t)$ and $V(y_t^*)$, respectively, so that for k large enough, $V(y_t^* \mid y_{t-k}^*)$ must exceed $V(y_t \mid y_{t-k})$, since the bound (7.7) holds for unconditional variances. However, for small k, Kleidon (1986b) shows that the inequality is reversed, i.e.

$$V(y_t^* \mid y_{t-k}^*) < V(y_t \mid y_{t-k})$$

and that this can hold for quite large k, depending on the size of ϕ: the closer ϕ is to 1, the greater the value of k must be before the above inequality is reversed. For example, if $\phi = 0.99$, then this inequality is only reversed when k is larger then 60.

Hence y_t^* *should* show greater smoothness than y_t, and such observed smoothness thus provides no evidence against either the variance bound (7.7) or the present value model (7.3)!

These difficulties in interpreting the inequality (7.7) are compounded by problems in estimating the unconditional variances making up the inequality. Since y_t^* will show less short-term variation than y_t then, even though their population unconditional means are both equal to $\mu_x/r(1-\phi)$, the variance of the sample mean of y_t^*, denoted $V(\bar{y}^*)$, will exceed that of y_t, $V(\bar{y})$, even in large samples. This is established by Kleidon (1986a, propositions 3–5) for the model in which x_t follows the AR(1) process (7.10), and has the consequence that estimates of sample variances based on sample means, rather than population means, are downward-biased. The bias is considerably more severe for y_t^*, so that the test of (7.7) is biased upward, i.e., that the inequality is violated far too often. This bias is exacerbated by imposing the terminal condition $y_T^* = y_T$. Although this arbitrary terminal condition becomes less important as T increases, and at the same time sample variances become better estimators of the unconditional variances used in (7.7), the small-sample bias can still be important in sample sizes typically used in empirical tests of the present value model: see also the analysis of Mankiw, Romer and Shapiro (1985).

A further problem involved in using variance bounds tests is rather more serious: as we have mentioned above, the inequalities (7.7) to (7.9) require that the variances of y_t, y_t^* and x_t be constant, i.e., that y_t and x_t be *stationary*. In typical applications this is unlikely to be the case and Shiller (1981a) attempts to induce stationarity by prior detrending using an exponential time trend. Not surprisingly, this has caused a great deal of controversy, for it implicitly assumes that both y_t and x_t are *trend stationary* processes (recall the discussion in chapter 3.1).

Kleidon (1986b) thus considers a non-stationary variant of (7.10) by setting $\phi = 1$, so that x_t, and hence y_t, follows a random walk and is thus *difference stationary*. In this situation, the conditional variances are

$$V(y_t \mid y_{t-k}) = \frac{\sigma_a^2 k}{r^2},$$

and

$$V(y_t^* \mid y_{t-k}) = \frac{\sigma_a^2}{r^2}\left[k + \frac{(1+r)^2}{2r+r^2}\right],$$

which show that the unconditional variances are not defined, so that the inequality (7.7) involves undefined terms, although the inequality (7.11) can be examined.

Kleidon (1986b) finds from a simulation exercise that using the variance inequality (7.7) when x_t and y_t are related by the present value model (7.3), but where x_t, and hence y_t, is non-stationary, almost always results in the violation of the bound, and often this violation can be 'gross', defined to be when $V(y_t) \geq 25V(y_t^*)$.

Mankiw, Romer and Shapiro (1985), henceforth MRS, propose an alternative volatility test which, they claim, avoids the problems encountered by both small-sample bias and inappropriate detrending. They consider a 'naive forecast' of y_t, given by

$$y_{t+i}^{\dagger} = \theta(1-\delta)\sum_{i=0}^{\infty}\delta^{i+1}x_{t+i}^{\dagger} + c,$$

where $x_{t+i}^{\dagger}$ denotes a naive forecast of x_{t+i} made at time t. This naive forecast need not be a rational one but it is assumed that rational agents at t have access to it. From the identity

$$y_t^* - y_t^{\dagger} = (y_t^* - y_t) + (y_t - y_t^{\dagger})$$

it follows that, since

$$E[(y_t^* - y_t)(y_t - y_t^{\dagger}) \mid \Phi_t] = E[u_t(y_t - y_t^{\dagger}) \mid \Phi_t] = 0,$$

we must have

$$E[(y_t^* - y_t^{\dagger}) \mid \Phi_t]^2 = E[(y_t^* - y_t) \mid \Phi_t]^2 + E[(y_t - y_t^{\dagger}) \mid \Phi_t]^2.$$

This implies that

$$E[(y_t^* - y_t^\dagger) \mid \Phi_t]^2 \geq E[(y_t^* - y_t) \mid \Phi_t]^2$$

and

$$E[(y_t^* - y_t^\dagger) \mid \Phi_t]^2 \geq E[(y_t - y_t^\dagger) \mid \Phi_t]^2.$$

These inequalities form the basis for MRS's tests, with the expectations being estimated as

$$T^{-1}\sum_{t=1}^{T}(y_t^* - y_t^\dagger)^2, \quad T^{-1}\sum_{t=1}^{T}(y_t^* - y_t)^2, \quad T^{-1}\sum_{t=1}^{T}(y_t - y_t^\dagger)^2,$$

respectively. Since these mean square errors do not use sample means in their construction, MRS claim that small-sample biases are avoided. Moreover, for suitable choices of the naive forecast $y_t^\dagger$, the random variables $y_t^* - y_t^\dagger$, $y_t^* - y_t$ and $y_t - y_t^\dagger$ will all be stationary even if x_t is generated by a process containing a unit root. MRS suggest setting $y_t^\dagger = (\delta/(1-\delta))x_{t-1}$, consistent with x_t following a random walk, while Mattey and Meese (1986) consider $y_t^\dagger = (1/\delta)y_{t-1} - x_{t-1}$, which is based on the present value recursion, and $y_t^\dagger = y_{t-1}$, which requires no unknown parameters in its construction.

Yet another volatility test that claims to avoid the problems encountered with the original tests has been proposed by West (1988b), who shows that, under the present value model (7.3),

$$E[y_{t+1} + x_{t+1} - E(y_{t+1} + x_{t+1} \mid x_t^0)]^2 \geq E[y_{t+1} + x_{t+1} - E(y_{t+1} + x_{t+1} \mid \Phi_t)]^2,$$

where, as in chapter 5, $x_t^0 = (x_t, x_{t-1}, \ldots, x_1)$, i.e., that the variance of revisions to forecasts of $y_{t+1} + x_{t+1}$ declines as the information set used for forecasting increases. Since market participants are presumed to use all the variables in Φ_t to forecast optimally, market forecasts thus tend to be more precise than forecasts based on a more limited information set: for example, that which uses just past values of x.

West's inequality can be made operational in the following way. The present value equation (7.4) can be written as

$$\begin{aligned} y_t &= \delta(y_{t+1} + x_{t+1}) - \delta[y_{t+1} + x_{t+1} - E(y_{t+1} + x_{t+1} \mid \Phi_t)] \\ &= \delta(y_{t+1} + x_{t+1}) + \epsilon_{t+1}, \end{aligned} \tag{7.14}$$

where the forecast error variance, σ_ϵ^2, is given by

$$\sigma_\epsilon^2 = \delta^2 E[y_{t+1} + x_{t+1} - E(y_{t+1} + x_{t+1} \mid \Phi_t)]^2.$$

Equation (7.14) can be estimated by IV techniques (see chapter 5.1), using as instruments variables known at time t: current and past values of x, for example. The resulting estimates, $\tilde{\delta}$ and $\tilde{\sigma}_\epsilon^2$, can then be used to estimate

$E[y_{t+1}+x_{t+1}-E(y_{t+1}+x_{t+1}|\Phi_t)]^2$ as $\tilde{\delta}^{-2}\tilde{\sigma}_\epsilon^2$. The left-hand side of the inequality can be obtained by assuming, for example, that $\nabla^d x_t$ follows an AR(p) process

$$\nabla^d x_t = \mu_x + \sum_{i=1}^{p} \phi_i \nabla^d x_{t-i} + a_t,$$

which can, of course, be estimated by LS to obtain estimates $\hat{\mu}_x, \hat{\phi}_1, \ldots, \hat{\phi}_p$ and $\hat{\sigma}_a^2$. $E[y_{t+1}+x_{t+1}-E(y_{t+1}+x_{t+1}|\Phi_t)]^2$ is then estimated as $\hat{\varphi}^2\hat{\sigma}_a^2$, where

$$\hat{\varphi} = (1-\tilde{\delta})^{-d}\left(1-\sum_{i=1}^{p}\tilde{\delta}^i\hat{\phi}_i\right)^{-1}$$

(see West, 1988b). Hence, the present value model implies that $\hat{\varphi}^2\hat{\sigma}_a^2 \geq \tilde{\delta}^{-2}\tilde{\sigma}_\epsilon^2$, a reversion of the inequality being taken as evidence in favour of excess volatility. A test of the hypothesis $H_0{:}f(\Theta) = \varphi^2\sigma_a^2 - \delta^{-2}\sigma_\epsilon^2 \geq 0$ exploits the asymptotic normality of the estimated parameter vector $\hat{\Theta} = (\tilde{\delta}, \hat{\mu}_x, \hat{\phi}_1, \ldots, \hat{\phi}_p, \tilde{\sigma}_\epsilon^2, \hat{\sigma}_a^2)$, with the standard error of $f(\Theta)$ being calculated as $(\partial f/\partial\Theta)\mathbf{V}(\partial f/\partial\Theta)'$, $\mathbf{V}$ being the covariance matrix of $\hat{\Theta}$: again see West (1988b) for details.

Example 7.2: Is there excess volatility in stock markets?

Since Shiller's (1981a) seminal article examining excess volatility in US stock prices, many articles have used his data set, or extensions of it, to assess the performance of the alternative volatility tests that have been discussed above: an accessible survey of this literature is provided by West (1988c), while a survey of the various econometric aspects of the volatility tests may be found in Gilles and LeRoy (1991).

Shiller's original data comprised annual observations on the real Standard and Poor's Composite Stock Price Index (y_t) and the associated real Dividend Index (x_t) from 1871 to 1979. In an attempt to detrend the two series, both were divided by a factor proportional to a common long-run exponential growth path. Using these detrended series and the perfect foresight series y_t^*, obtained by recursively solving $y_t^* = \delta(y_{t+1}^* + x_t)$ backwards from the terminal condition $y_{1979}^* = y_{1979}$, Shiller (1981b) reports the following statistics

$$\hat{\sigma}(y) = 47.2, \quad \hat{\sigma}(\nabla y) = 24.3, \quad \hat{\sigma}(y^*) = 7.51$$
$$\hat{\sigma}(x) = 1.28, \quad \hat{\sigma}(\nabla x) = 0.768.$$

Inequality (7.7) is seen to be grossly violated. With the discount rate r estimated as $\bar{x}/\bar{y} = 0.0452$, (7.8) implies that an upper bound to $\hat{\sigma}(\nabla y)$ is 4.26: this is seen to be grossly violated as well. Similarly, (7.9) implies an upper bound to $\hat{\sigma}(\nabla y)$ of 59.0: this bound is *not* violated.

Subsequent research has focused on the gross violation of inequality (7.7). Kleidon (1986a), in assessing the importance of small-sample bias in

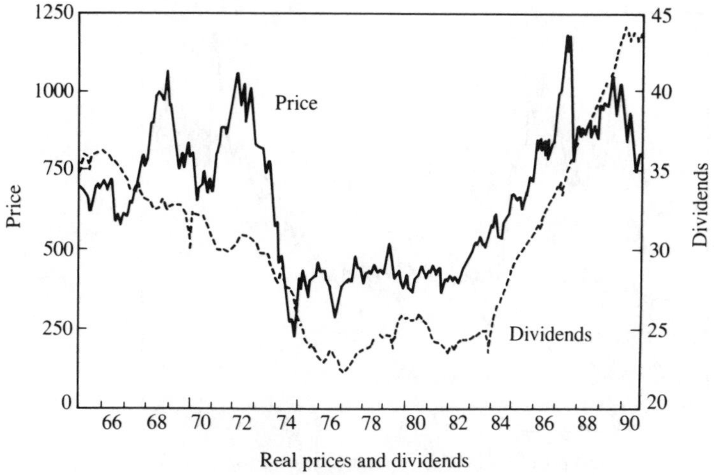

Figure 7.1 *FTA All Share* index (monthly 1965–90)

these estimates, concludes that such bias, on its own, does not seem sufficient to account for a violation of the bound of this magnitude.

Kleidon (1986b) presents evidence from unit root tests that both prices and dividends are $I(1)$ processes and hence Shiller's detrending procedure may thus be regarded as problematic. He then provides evidence that conditional variance inequalities based on (7.11) are not violated when using a subsample (1926–79) of Shiller's data.

Mankiw, Romer and Shapiro (1985) extend the sample period to 1983 and compute their inequalities using the naive forecast $y_t^\dagger = (\delta/(1-\delta))x_{t-1} = r^{-1}x_{t-1}$ for a variety of r values in the range $0.04 \leq r \leq 0.10$, finding that in every case the inequalities are violated. They also repeat the computations with an adjustment for heteroskedasticity, obtaining similar, although somewhat weaker, violations.

West (1988b) finds, for various specifications of the ARIMA process generating x_t, that the computed value of $\varphi^2\sigma_a^2 - \delta^{-2}\sigma_\epsilon^2$ is always significantly negative, thus providing further evidence in favour of excess volatility.

The general impression from these tests is that, although the original Shiller results were subject to problems of small-sample bias and non-stationarity that may have affected the magnitude of his findings of excessive volatility, when these difficulties are tackled excess volatility is still found – albeit of an order of magnitude smaller than first obtained.

Is there evidence of excess volatility in UK stock prices? We investigate this question by considering monthly observations from January 1965 to December 1990 on real prices and dividends obtained by dividing the *FTA All Share* price and dividend series by the retail price index. Figure 7.1

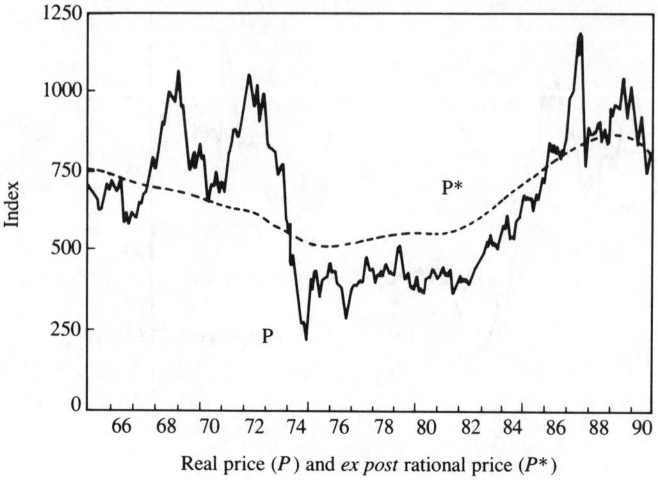

Figure 7.2 *FTA All Share* index (monthly 1965–90)

provides plots of the resulting series. In implementing Shiller's original inequalities, we note first that neither of the series shows any steady upward trend and hence detrending by dividing by an exponential growth factor would seem to be inappropriate. We thus analyse the observed series, from which we estimate r as 0.0460, this being the ratio of the sample means. Using this value, y_t^* is then calculated in the usual fashion, leading to the following statistics

$$\hat{\sigma}(y)=220.0, \quad \hat{\sigma}(\nabla y)=40.7, \quad \hat{\sigma}(y^*)=109.7$$
$$\hat{\sigma}(x)=30.2, \quad \hat{\sigma}(\nabla x)=0.39.$$

Inequality (7.7) is thus violated, as can clearly be seen in figure 7.2, which plots y_t and y_t^* and produces a picture that has very similar characteristics to those presented by Shiller (1981a, 1981b) for the US. The other two inequalities are also violated, the implied upper bounds for $\sigma(\nabla y)$ being 18.5 and 29.2, respectively.

Of course, use of these inequalities is only justified if prices and dividends are stationary, and visual examination of the series in figure 7.1 suggests that this is unlikely to be the case. Formal unit root tests confirm this, and further analysis reveals that both y_t and x_t can be adequately modelled as driftless random walks. As Kleidon's (1986b) analysis demonstrates, in such circumstances the appropriate inequality to consider is that between the conditional variances, i.e., (7.11). Since the conditional means in this case are simply $E(y_t \mid y_{t-k}) = E(y_t^* \mid y_{t-k}) = y_{t-k}$, the conditional variances can be estimated as

Table 7.1 *Estimates of the ratio of the conditional variances of* y *and* y*

k	$V(y_t \mid y_{t-k})/V(y_t^* \mid y_{t-k})$
1	0.07
2	0.14
3	0.21
6	0.41
12	0.79
18	1.07
24	1.21
48	1.23
96	1.49

$$(T-k)^{-1}\sum_{t=k+1}^{T}(y_t-y_{t-k})^2 \text{ and } (T-k)^{-1}\sum_{t=k+1}^{T}(y_t^*-y_{t-k})^2$$

respectively. Estimates for various values of k are presented in table 7.1, where it is seen that for small values of k the inequality (7.11) is *not* violated, although from $k=18$ onwards it is violated by an increasing margin.

The Mankiw, Romer and Shapiro (1985) MRS test with the naive forecast given by $y_t^{\dagger}=r^{-1}x_{t-1}=21.73x_{t-1}$ leads to the following estimates

$$(T-1)^{-1}\sum_{t=2}^{T}(y_t^*-y_t^{\dagger})^2=3552, \quad (T-1)^{-1}\sum_{t=2}^{T}(y_t^*-y_t)^2=24542,$$

$$(T-1)^{-1}\sum_{t=2}^{T}(y_t-y_t^{\dagger})^2=22435,$$

so that both inequalities making up the MRS test are violated. However, using Mattey and Meese's (1986) suggestions for $y_t^{\dagger}$ does not lead to a violation: indeed, their second choice of $y_t^{\dagger}=y_{t-1}$ is equivalent to $k=1$ in table 7.1, while their first choice of $y_t^{\dagger}=(1+r)y_{t-1}-x_{t-1}=1.046y_{t-1}-x_{t-1}$ is very similar, differing from it only by the amount $r(y_{t-1}-x_{t-1})=\bar{x}((y_{t-1}/\bar{y})-(x_{t-1}/\bar{x}))$, which will typically be rather small.

To compute West's (1988b) test, we first run the IV regression of y_t on $(y_{t+1}+x_{t+1})$, using x_t and x_{t-1} as instruments, to obtain the estimates $\tilde{\delta}=0.957$ and $\hat{\sigma}_{\epsilon}^2=40.1^2$. Since x_t is a driftless random walk, $\hat{\varphi}^2\hat{\sigma}_a^2=(1-\tilde{\delta})^{-2}\hat{\sigma}^2(\nabla x)$ and hence the statistic $\varphi^2\sigma_a^2-\delta^{-2}\sigma_{\epsilon}^2$ is calculated to be $(0.39/0.043)^2-(40.1/0.957)^2=82.3-1755.8=-1673.5$. Although we do not provide a formal test of the hypothesis that $\varphi^2\sigma_a^2-\delta^{-2}\sigma_{\epsilon}^2$ is non-negative, we regard the sign and size of the statistic as providing further evidence of there being excess volatility in UK stock prices.

7.2.3 Cointegration and tests of present value models

Campbell and Shiller (1987, 1988a) propose an alternative approach to testing the present value model which explicitly takes into account non-stationarity of x_t and y_t, linking the implications of the model to the concept of cointegration discussed in chapter 6.

Consider subtracting $(\delta/(1-\delta))x_t$ from both sides of (7.3). On rearranging, we obtain a new variable, S_t, which Campbell and Shiller (1987) term the 'spread'

$$S_t = y_t - \theta x_t = \theta \sum_{i=1}^{\infty} \delta^i E(\nabla x_{t+i} | \Phi_t). \tag{7.15}$$

Suppose y_t and x_t are both $I(1)$ processes. It then follows from (7.15) that S_t must be $I(0)$, which in turn implies that y_t and x_t are cointegrated with cointegrating parameter θ.

If y_t and x_t are cointegrated $I(1)$ variables, then S_t and ∇x_t must together form a jointly covariance stationary process. For a suitably chosen p, this can be approximated in finite samples by the VAR

$$\begin{aligned} S_t &= \sum_{i=1}^{p} a_i S_{t-i} + \sum_{i=1}^{p} b_i \nabla x_{t-i} + \epsilon_t \\ \nabla x_t &= \sum_{i=1}^{p} c_i S_{t-i} + \sum_{i=1}^{p} e_i \nabla x_{t-i} + \eta_t \end{aligned} \tag{7.16}$$

which can be expressed in *companion form* as

$$\begin{bmatrix} S_t \\ \cdot \\ \cdot \\ \cdot \\ S_{t-p+1} \\ \nabla x_t \\ \cdot \\ \cdot \\ \cdot \\ \nabla x_{t-p+1} \end{bmatrix} = \begin{bmatrix} a_1 & \cdots & a_p & b_1 & \cdots & b_p \\ 1 & & & & & \\ & \ddots & & & & \\ & 1 & 0 & & & \\ c_1 & \cdots & c_p & e_1 & \cdots & e_p \\ & & & 1 & & \\ & & & & \ddots & \\ & & & & 1 & 0 \end{bmatrix} \begin{bmatrix} S_{t-1} \\ \cdot \\ \cdot \\ \cdot \\ S_{t-p} \\ \nabla x_{t-1} \\ \cdot \\ \cdot \\ \cdot \\ \nabla x_{t-p} \end{bmatrix} + \begin{bmatrix} \epsilon_t \\ 0 \\ \vdots \\ 0 \\ \eta_t \\ 0 \\ \vdots \\ 0 \end{bmatrix},$$

where blank elements are zero. This can be written more compactly, in an obvious notation, as

$$\mathbf{z}_t = \mathbf{A}\mathbf{z}_{t-1} + \mathbf{v}_t. \tag{7.17}$$

We can use the first-order formulation (7.17) to express the variant of the present value model presented as equation (7.15) in closed-form solution, i.e., as a function of variables known to agents at the time expectations are formed. If we denote the restricted information set $\mathbf{H}_t$ as consisting only of current and lagged S_t and ∇x_t, i.e., $\mathbf{H}_t = (S_t^0, \nabla x_t^0)$, then the expectation of future values of $\mathbf{z}_t$, conditional on $\mathbf{H}_t$, is

$$E(\mathbf{z}_{t+i} \mid \mathbf{H}_t) = \mathbf{A}^i \mathbf{z}_t . \tag{7.18}$$

Define $\mathbf{g}$ as a $(2p \times 1)$ selection vector with unity as the first element and zeros elsewhere, and $\mathbf{h}$ as another selection vector with unity as the $(p+1)$-th element and zeros elsewhere. Then, using (7.18), equation (7.15) can be projected on to the restricted information set $\mathbf{H}_t$ and, using the law of iterated projections, we may write

$$\mathbf{g}'\mathbf{z}_t = \theta \mathbf{h}' \delta \mathbf{A}(\mathbf{I} - \delta \mathbf{A})^{-1} \mathbf{z}_t , \tag{7.19}$$

which is a closed-form variant of the present value model. The advantage of this formulation is that it imposes the model's restrictions on the coefficients of the VAR, since if (7.19) is to hold non-trivially, the following $2p$ restrictions must hold

$$\mathbf{g}' - \theta \mathbf{h}' \delta \mathbf{A}(\mathbf{I} - \delta \mathbf{A})^{-1} = 0 .$$

Although these restrictions appear complex and hard to interpret, for a given δ, and hence θ, they turn out to be equivalent to the following set of linear restrictions

$$\begin{aligned} &1 - \delta a_1 - \theta \delta c_1 = 0 \\ &a_i + \theta c_i = 0, \quad i = 2, \ldots, p \\ &b_i + \theta e_i = 0, \quad i = 1, \ldots, p . \end{aligned} \tag{7.20}$$

These restrictions can be interpreted as follows. The present value model implies

$$E(y_t - \delta^{-1} y_{t-1} + x_{t-1} \mid \mathbf{H}_{t-1}) = 0 .$$

In terms of S_t and ∇x_t, this can be written as

$$E(S_t - \delta^{-1} S_{t-1} + \theta \nabla x_t \mid \mathbf{H}_{t-1}) = 0 . \tag{7.21}$$

Using the VAR formulation (7.16), we have

$$E(S_t - \delta^{-1} S_{t-1} + \theta \nabla x_t \mid \mathbf{H}_{t-1}) = \sum_{i=1}^{p} (a_i + \theta c_i) S_{t-i} + \sum_{i=1}^{p} (b_i + \theta e_i) \nabla x_{t-i},$$

which is identically equal to zero under the restrictions (7.20). Thus the restrictions can be related directly back to the proposition, implied by the present value model, that $\delta^{-1} y_t + x_t$ should be an optimal predictor of y_{t+1}, cf. equation (7.14).

A further implication of the present value model for the VAR (7.17) is that S_t must Granger-cause ∇x_t unless S_t is itself an *exact* linear function of (x_t^0). This is because S_t is an optimal forecast of a weighted sum of future values of ∇x_t conditional on Φ_t (recall equation (7.15)). S_t will therefore have incremental explanatory power for future ∇x_t if agents have information useful for forecasting ∇x_t beyond (x_t^0): if not, they form S_t as an exact linear function of (x_t^0).

Following Campbell and Shiller (1987), we can also use these restrictions to construct volatility-based tests of the model. If we define the 'theoretical spread', S_t^*, as

$$S_t^* = \theta \sum_{i=1}^{\infty} \delta^i E(\nabla x_{t+i} \mid \mathbf{H}_t) = \theta \mathbf{h}' \delta \mathbf{A}(\mathbf{I} - \delta \mathbf{A})^{-1} \mathbf{z}_t,$$

then, if the present value model is correct, we have from (7.15)

$$S_t^* = S_t,$$

and hence

$$V(S_t^*) = V(S_t).$$

This equality thus provides a way of assessing the model informally by examining the comovement of $V(S_t^*)$ and $V(S_t)$. In particular, if the model is correct, the ratio $V(S_t)/V(S_t^*)$ should differ from unity only because of sampling error in the estimated coefficients of the VAR.

Campbell and Shiller (1987) also suggest a second volatility test in addition to this 'levels variance ratio'. Denoting the innovation in (7.21) as ξ_t, i.e.

$$\xi_t = S_t - \delta^{-1} S_{t-1} + \theta \nabla x_t,$$

the 'theoretical innovation', ξ_t^*, can be analogously defined as

$$\xi_t^* = S_t^* - \delta^{-1} S_{t-1}^* + \theta \nabla x_t.$$

Under the present value model, $\xi_t^* = \xi_t$ since $S_t^* = S_t$, so that the 'innovation variance ratio', $V(\xi_t)/V(\xi_t^*)$, should again be compared with unity.

A variance bounds test more closely related to those discussed previously can be constructed as follows. From (7.15) we have

$$S_t = E(\tilde{S}_t \mid \Phi_t),$$

where

$$\tilde{S}_t = \theta \sum_{i=1}^{\infty} \delta^i \nabla x_{t+i}.$$

It is easily seen that $\tilde{S}_t$ follows the recursive formula

$$\tilde{S}_t = \delta(\theta \nabla x_{t+1} + \delta \tilde{S}_{t+1}). \qquad (7.22)$$

If, following Campbell and Shiller (1987), we impose the terminal condition $S_T = \tilde{S}_T$, where T is the final point in the sample, then (7.22) can be used to construct a series for $\tilde{S}_t$. Since the realisation of $\tilde{S}_t$ must be equal to its rationally expected value plus a forecast error orthogonal to the expected value, relation (7.22) implies the variance inequality $V(\tilde{S}_t) > V(S_t)$.

Example 7.3: The cointegration approach to testing stock market volatility

Campbell and Shiller (1987) employ this 'cointegration approach' to test the Standard and Poor data extended to 1986 and incorporating some corrections to the dividend series. As a preliminary, unit root tests are needed to ensure that both y_t and x_t are $I(1)$: this was indeed found to be the case. Less conclusive evidence was presented that the spread was stationary, which would imply that y_t and x_t are cointegrated (in practice, Campbell and Shiller use $SL_t = y_t - \theta x_{t-1}$ rather than S_t to avoid timing problems caused by the use of beginning-of-year stock prices and dividends paid within-year).

Nonetheless, assuming cointegration and using the cointegrating regression estimate of θ (an implied discount rate of 3.2 per cent), a second-order VAR was constructed for the bivariate $(SL_t, \nabla x_t)$ process. The estimates suggested that dividend changes were highly predictable and there was strong evidence that the spread Granger-caused dividend changes, one implication of the present value model. The restrictions (7.20) could not be rejected at conventional significance levels, and neither were the two variance ratios significantly larger than unity.

Markedly different results were obtained, however, when the sample mean return was used to calculate a discount rate of 8.2 per cent. Now the restrictions (7.20) could be rejected at low significance levels and the two variance inequalities were sharply violated. Campbell and Shiller suggest that the implied discount rate of 3.2 per cent obtained from the cointegrating regression may be too low: nevertheless, although they prefer to use the higher discount rate of 8.2 per cent, which leads to the finding of excessive volatility, they do emphasise that the strength of the evidence depends sensitively on the assumed value of the discount rate.

Mills (1992a) has applied this technique to the UK data used in example 7.2. We already know that both prices and dividends are $I(1)$ and it is also the case that they are cointegrated, so that the spread is $I(0)$: for example, the cointegrating regression of y_t on x_t yields a θ estimate of 30.01 (i.e., a discount rate of 3.3 per cent) and a unit root test statistic for the residuals of -2.98, which is able to reject the null of non-cointegration at the 12.5 per cent level of significance. Johansen's (1988) ML approach provides more conclusive evidence in favour of cointegration: it yields the estimate

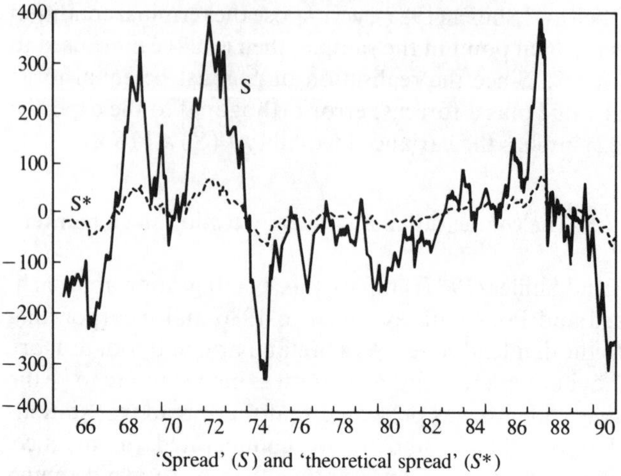

Figure 7.3 *FTA All Share* index: 'excess volatility' (monthly 1965–90)

$\hat{\theta}=23.36$ (a discount rate of 4.3 per cent) and is able to reject non-cointegration at significance levels of less than 1 per cent.

Use of both θ estimates led to fourth-order VARs being fitted, from which rejections of the restrictions (7.20) at very low significance levels (less than 1 per cent) were obtained. In both cases the variance ratio inequalities were also sharply violated: for example, using $\hat{\theta}=30.01$ from the cointegrating regression obtains $V(S_t)/V(S_t^*)=29$ and $V(\xi_t)/V(\xi_t^*)=8.5$. The implied 'excess volatility' can readily be seen in figure 7.3 which plots S_t and S_t^* using this estimate of θ: when the ML estimate is used, this excess volatility is even more marked, for the variance ratios are almost twice as large and the theoretical spread is considerably smoother than the series shown in the figure.

Example 7.4: Testing the expectations hypothesis of the term structure of interest rates

Shiller (1979) shows that the expectations hypothesis of the term structure of interest rates, that the current long interest rate is the weighted average of current and expected future short rates, can be put into the form of the present value model (7.3). In this framework, y_t is the current interest rate (the yield to maturity) on a long bond (strictly a perpetuity), x_t is the current one-period interest rate, θ is set to unity, δ is a parameter of linearisation, typically set equal to $(1+\bar{y})^{-1}$, and c is a liquidity premium unrestricted by the model.

The expectations hypothesis thus asserts that, if y_t and x_t are both $I(1)$, then the spread, $S_t=y_t-x_t$ (noting that $\theta=1$), must be $I(0)$ and hence that y_t

and x_t must be cointegrated with cointegrating vector $(1, -1)$. S_t and ∇x_t then have the VAR representation (7.16) and the expectations hypothesis implies the restrictions given by equation (7.20). A further implication of the model is that the excess return on holding a long bond for one period rather than a one-period bond should be unpredictable.

Although this form of the expectations hypothesis is only strictly valid when the long rate is a perpetuity, it can still be used for bonds of finite, but very long, life, e.g., twenty years. Campbell and Shiller (1987) thus test the model using monthly data on the yield on US Treasury twenty-year bonds and one-month Treasury bill rates for the period 1959 to 1983. Evidence is presented confirming that both series are $I(1)$ and that the spread is stationary, but a test of the restrictions (7.20) rejects the expectations hypothesis very strongly. Nevertheless, the variance ratios are not significantly different from unity and the 'theoretical spread', S_t^*, is highly correlated with the actual spread, S_t. Campbell and Shiller interpret these conflicting findings as evidence that deviations from the present value model are only transitory and suggest that the model does, in fact, fit the data comparatively well.

Further support for the expectations hypothesis using these techniques has been provided from UK data by MacDonald and Speight (1988) and Mills (1991c). In this latter paper, quarterly observations for the sample periods 1870 to 1913, 1919 to 1939, and 1952 to 1988 (plus various subperiods) are investigated using a variety of interest rates. For all interest rates in all sample periods, the hypothesis that they are $I(1)$ cannot be rejected: similarly, all spreads are found to be $I(0)$. For the pre-Second World War data, the restrictions (7.20) are rejected at very small significance levels for all subperiods, and all variance ratios exceed unity, usually by a large margin.

A very different pattern emerges from analysis of the post-Second World War data where, typically, the expectations hypothesis is not rejected either by testing the implied restrictions or by computing variance ratios. This is particularly so for data from the 1980s and suggests that the evidence in favour of using the present value model (7.3) to analyse the setting of interest rates is surprisingly strong.

7.3 Rational bubbles

7.3.1 The theory of rational bubbles

The volatility tests developed in section 2 have all been based on the assumption that the transversality condition in equation (7.5) holds, i.e., that

$$\lim_{n\to\infty} \delta^n E(y_{t+n} \mid \Phi_t) = 0 .$$

If this is the case, then $y_t = y_t^f$, where y_t^f is the unique forward solution, often termed the 'market fundamentals' solution

$$y_t^f = \sum_{i=0}^{\infty} \delta^{i+1} E(x_{t+i} \mid \Phi_t).$$

If this transversality condition fails to hold, however, there will be a family of solutions to (7.5): see, for example, Blanchard and Watson (1982), West (1987), and Diba and Grossman (1987, 1988). In such circumstances, any y_t that satisfies

$$y_t = y_t^f + B_t, \tag{7.23}$$

where

$$E(B_{t+1} \mid \Phi_t) = \delta^{-1} B_t = (1+r) B_t \tag{7.24}$$

is also a solution. B_t is known as a *speculative*, or *rational*, *bubble*, an otherwise extraneous event that affects y_t because everyone expects it to do so, i.e., it is a self-fulfilling expectation. An example of such a bubble is (see Blanchard and Watson, 1982, and West, 1987)

$$B_t = \begin{cases} (B_{t-1} - \bar{B})/\pi\delta & \text{with probability } \pi \\ \bar{B}/(1-\pi)\delta & \text{with probability } 1-\pi \end{cases}, \tag{7.25}$$

where $0 < \pi < 1$ and $\bar{B} > 0$ (other examples are provided by, for example, Hamilton, 1986).

According to (7.25), strictly positive bubbles grow and burst, with the probability that the bubble bursts being $1-\pi$. While the bubble floats it grows at a rate $(\delta\pi)^{-1} = (1+r)/\pi > 1+r$: investors in the asset thus receive an extraordinary return to compensate them for the capital loss that would have occurred had the bubble burst.

Equation (7.23) implies that the rational bubble has explosive conditional expectations, since

$$E(B_{t+i} \mid \Phi_t) = (1+r)^i B_t$$

and $r > 0$. Thus, if y_t is the price of a freely disposable asset, say a stock, then a negative rational bubble ($B_t < 0$) cannot exist, because its existence would imply that y_t decreases without bound at the geometric rate $(1+r)$, so that it becomes negative at some finite time $t+i$. Negative rational bubbles are, at least theoretically, possible if y_t is an exchange rate, for this characterises a continual currency appreciation.

While positive bubbles are theoretically possible, Diba and Grossman (1987, 1988) discuss a number of conditions that must be met for their existence. Positive bubbles imply that asset holders might expect such a bubble to come to dominate y_t, which would then bear little relation to market fundamentals. Bubbles would only be empirically plausible if, despite explosive conditional expectations, the probability is small that a rational bubble becomes arbitrarily large. Moreover, for exchange rates a positive rational bubble would imply a continual devaluation of the currency, and this can be ruled out by an argument symmetric to that used above for a negative rational bubble in stock prices.

Diba and Grossman also show that, if a rational bubble does not exist at time t, then it cannot get started at any later date $t+i$, $i>0$, and that if an existing rational bubble bursts, a new independent rational bubble cannot simultaneously start. Thus, if a rational bubble exists at time t, then it must have started at time $t=0$ (the first date of trading of the asset), it has not burst, it will not restart if it bursts, and, if it is a bubble in a stock price, the stock has been continuously overvalued relative to market fundamentals.

7.3.2 Testing for rational bubbles

West (1987) develops a test for rational bubbles based on his approach to volatility testing discussed in the previous section. The idea behind this test is as follows. Suppose that x_t is generated by the AR(1) process of equation (7.10)

$$x_t=\mu_x+\phi x_{t-1}+a_t. \qquad (7.26)$$

As we have shown, this implies that, if $y_t=y_t^f$

$$y_t=\gamma+\alpha x_t, \qquad (7.27)$$

where $\alpha=\phi/(1+r-\phi)=\phi\delta/(1-\phi\delta)$ and $\gamma=\mu_x\alpha/\delta r\phi=\mu_x\delta/(1-\delta)(1-\phi\delta)$. LS estimation of equations (7.26) and (7.27) will then provide consistent estimates $\hat{\mu}_x$, $\hat{\phi}$, $\hat{\gamma}$ and $\hat{\alpha}$. If there is a bubble, however, y_t will not be equal to y_t^f, and $\hat{\gamma}$ and $\hat{\alpha}$ will be inconsistent. Even when there is a bubble, however, a consistent estimate of δ is provided by the IV estimate $\tilde{\delta}$ obtained from (7.14). West's specification test is thus to compare the two implied estimates of γ and α, i.e., to test

$$H_0\colon \hat{\gamma}=\frac{\tilde{\delta}\hat{\mu}_x}{(1-\tilde{\delta})(1-\hat{\phi}\tilde{\delta})}, \qquad \hat{\alpha}=\frac{\hat{\phi}\tilde{\delta}}{(1-\hat{\phi}\tilde{\delta})},$$

rejecting the no bubbles null for large values of the test statistic. If the null hypothesis is denoted $R(\theta)=0$, where $\theta=(\mu_x,\phi,\gamma,\delta)$, then the test statistic takes the form

$$R(\hat{\theta})'\left[\left(\frac{\partial R}{\partial \hat{\theta}}\right)\mathbf{V}\left(\frac{\partial R}{\partial \hat{\theta}}\right)'\right]^{-1}R(\hat{\theta})\sim\chi_2^2,$$

where $\mathbf{V}$ is the asymptotic covariance matrix of θ. West (1987) provides computational details and also considers extensions to the cases where x_t is generated by either an AR(p) process or an ARIMA(p,1,0) process. He also emphasises that a rejection of the null hypothesis could occur for reasons other than the existence of a bubble: for example, time-varying discount rates, and in this case he develops an extension of the test. West also argues that the bubbles explanation for the rejection of H_0 becomes more compelling if the constituent equations of the test, (7.14), (7.26) and (7.27), are known to be correctly specified: he thus advocates applying a battery of diagnostic tests to ensure this.

Bubbles can also be tested by examining their implications for cointegration between various series. When $y_t = y_t^f$, so no bubbles are present, equation (7.5) implies that

$$U_{t+1} = y_t - \delta(y_{t+1} + x_t)$$

must be $I(0)$ and, as we have already shown, the spread $S_t = y_t - \theta x_t$ must also be $I(0)$, so that y_t must be cointegrated with both x_t and $y_{t+1} + x_t$ (it must also be the case that ∇y_t is $I(0)$). If, on the other hand, a bubble is present, so that $y_t = y_t^f + B_t$, the bubble must appear in both U_t and S_t. Since, by definition, B_t is non-stationary, these variables cannot be $I(0)$ and the cointegration relationships cannot hold.

Hamilton and Whiteman (1985) discuss these implications in more detail, showing that, if $x_t \sim I(d)$, then rational bubbles can only exist if $y_t \sim I(d+b)$, where $b > 0$. However, the finding that y_t is of a higher order of integration than x_t is not necessarily evidence in favour of bubbles. As Hamilton and Whiteman point out, such a finding might be explained by numerous other factors: what appears to be a bubble could have arisen instead from rational agents responding solely to fundamentals not observed by the modeller.

One further important drawback with tests of stationarity and cointegration is the question of power. Diba (1990), for example, argues that if $\bar{B}$ in (7.25) is sufficiently close to zero, the ensuing bubble would generate fluctuations in a finite sample that could not be distinguished from stationary behaviour. Meese (1986) provides both simulation and empirical evidence on exchange rate bubbles that is consistent with this.

Example 7.5: Are there bubbles in stock market prices?

Given the strong evidence of excessive volatility in stock prices presented in examples 7.2 and 7.3, the obvious question to ask is whether such volatility is a consequence of a rational bubble being present or whether the rejection of the present value model is simply due to excess returns being predictable, perhaps because of time-varying discount rates.

The easiest way of determining whether a bubble is present is to examine the orders of integration of prices, dividends, the spread and the variable U. Campbell and Shiller (1987) report various unit root tests and conclude that, because ∇x_t and S_t are stationary, a rational bubble does not appear to be present in the US stock price data. Mills (1992a) examines this question for the UK data using the Dickey and Pantula (1987) procedure, discussed in chapter 3.1.4, for testing the possible presence of more than one unit root in prices and dividends. No evidence is found for the presence of bubbles.

As has been discussed above, such tests may have low power in detecting bubbles. This is, of course, just a reflection of the power problems that can bedevil unit root tests in any context and has been discussed in chapter 3. As a consequence, West (1987) has applied his test to the Standard and Poor data, finding that the null hypothesis of no bubbles is generally rejected. However, when time-varying discount rates are allowed for, the results are rather more ambiguous, with less evidence against the no bubble null being found.

7.4 The 'dividend-ratio model': a log-linear approximation to the present value model

When analysing the present value model of stock prices, an alternative way of incorporating non-stationarity into prices and dividends is to specify equation (7.3) in terms of the *price/dividend ratio* and *dividend growth rates*, i.e., if we now let P_t be real stock prices and D_t real dividends, so that $\eta_t = D_t/D_{t-1}$ represents dividend growth, we have

$$P_t/D_t = \sum_{i=0}^{\infty} \delta^{i+1} E(D_{t+i}/D_t \mid \Phi_t) = \delta + \sum_{i=1}^{\infty} \delta E\left(\left(\prod_{j=0}^{i} \delta\eta_{t+j}\right) \mid \Phi_t\right). \tag{7.28}$$

Cochrane and Sbordone (1988), for example, prove that P_t/D_t will be stationary if η_t is stationary, and this will also be true for time-varying discount rates as long as these are stationary as well and dividends do not grow 'too' fast. As Cochrane (1991b) notes, variance bounds can then be constructed using (7.28) rather than (7.3) without running into the non-stationarity problems discussed above.

Campbell and Shiller (1988c) focus attention on the logarithmic gross return. Recall the definition of the one-period return as, in this case

$$r_{t+1} = (P_{t+1} + D_t - P_t)/P_t = (P_{t+1} + D_t/P_t) - 1\,.$$

Taking logarithms and using the approximation $r_{t+1} \cong \ln(1 + r_{t+1}) = h_{1,t+1}$ yields

$$h_{1,t+1} = \ln(P_{t+1} + D_t) - \ln(P_t)\,.$$

Campbell and Shiller examine the relationship between $h_{1,t+1}$ and the logarithms of dividends and prices, $d_t = \ln(D_t)$ and $p_t = \ln(P_t)$, respectively. The relationship is non-linear, of course, but can be approximated as

$$h_{1,t+1} = k + \rho p_{t+1} + (1-\rho)d_t - p_t = k + \zeta_t - \rho\zeta_{t-1} + \nabla d_t, \tag{7.29}$$

where $\zeta_t = d_{t-1} - p_t$ is the logarithmic 'dividend/price ratio', ρ is the average of the ratio $P_t/(P_t + D_{t-1})$, which can be estimated as $\exp(\overline{\nabla d - h})$, and k is a constant given by $k = -\ln(\rho) - (1-\rho)\ln(1/\rho - 1)$: see Campbell and Shiller (1988b, 1988c) for details of the derivation of equation (7.29).

Equation (7.29) can be thought of as a difference equation relating ζ_t to ζ_{t+1}, ∇d_t and $h_{1,t+1}$ and, on solving forwards and imposing the terminal condition that $\lim_{i\to\infty}\rho^i\zeta_{t+i} = 0$, we obtain

$$\zeta_t \cong \sum_{i=0}^{\infty} \rho^i (h_{1,t+i+1} - \nabla d_{t+i}) - \frac{k}{1-\rho}. \tag{7.30}$$

As it stands, this equation has no economic content, since it simply says that ζ_t, the log dividend–price ratio, can be written as a discounted value of the differences between future returns and dividend growth rates discounted at the constant rate ρ, less a constant $k/(1-\rho)$.

Suppose, however, that, as before, expected one-period returns are constant: $E(h_{1,t+1} \mid \Phi_t) = r$. Then, on taking conditional expectations of (7.30), we obtain

$$\zeta_t \cong -\sum_{i=0}^{\infty} \rho^i E(\nabla d_{t+i} \mid \Phi_t) + \frac{r-k}{1-\rho}, \tag{7.31}$$

which expresses the log dividend-price ratio as a linear function of expected real dividend growth into the infinite future. The restrictions implicit in (7.31) can be tested using an analogous framework to that developed in section 7.2.3 above, noting that, in this context, ζ_t is the logarithmic counterpart of the spread $S_t = P_t - \theta D_t$. Thus we consider ζ_t and ∇d_t to be generated by a pth-order VAR, which can be written in companion form as in equation (7.17) with $\mathbf{z}_t = (\zeta_t, \nabla d_t)'$. The implied solution to the present value model conditional on the restricted information set $\mathbf{H}_t = (\zeta_t^0, \nabla d_t^0)$ is then

$$\mathbf{g}'\mathbf{z}_t = -\mathbf{h}'\mathbf{A}(\mathbf{I} - \rho\mathbf{A})^{-1}\mathbf{z}_t,$$

with the accompanying set of restrictions

$$\mathbf{g}' + \mathbf{h}'\mathbf{A}(\mathbf{I} - \rho\mathbf{A})^{-1} = 0. \tag{7.32}$$

As with the analogous set (7.20), these restrictions imply that $E(h_{1,t+1} \mid \mathbf{H}_t) = 0$, so that returns are unpredictable. Moreover, as with the VAR of (7.17), a further implication of this model is that ζ_t should Granger-cause ∇d_t.

Campbell and Shiller (1988c) argue that working with logarithms has certain advantages over the approach developed previously when testing the implications of the present value model for stock prices. One advantage is that it is easy to combine with individual log-linear models of prices and dividends which, as stressed by Kleidon (1986b), for example, are both more appealing on theoretical grounds and do appear to fit the data better than linear ones. A second advantage is that using the variables ζ_t and ∇d_t mitigates measurement error problems that may occur when deflating nominal stock prices and dividends by some price index to obtain real variables.

The model has been extended in various ways. Campbell and Shiller (1988c) allow expected log returns to be given by the model $E(h_{1,t+1} \mid \Phi_t) = r + R_t$, where R_t is the real return on, for example, Treasury bills. In this case $R_{t+i} - \nabla d_{t+i}$ replaces $-\nabla d_{t+i}$ in equation (7.31) and $\mathbf{z}_t = (\zeta_t, R_{t+i} - \nabla d_{t+i})$ becomes the vector modelled as a VAR. Campbell and Shiller (1988b) focus attention on the j-period discounted return

$$h_{jt} = \sum_{k=0}^{j-1} \rho^j h_{1,t+k},$$

which leads to the following set of restrictions on the VAR:

$$\mathbf{g}'(\mathbf{I} - \rho^j \mathbf{A}^j) + \mathbf{h}'\mathbf{A}(\mathbf{I} - \rho\mathbf{A})^{-1}(\mathbf{I} - \rho^j \mathbf{A}^j) = 0.$$

Although these restrictions are *algebraically* equivalent to those of (7.32) for all j, reflecting the fact that if one-period returns are unpredictable, then j-period returns must also be, and vice-versa, Wald tests may yield different results depending on which value of j is chosen. Nevertheless, the VAR framework confers yet another advantage in this setup: it needs to be estimated only once, as tests can be conducted for any j without reestimating the system.

Campbell and Shiller (1988b) also extend the VAR framework to incorporate a third variable, a long moving average of the earnings–price ratio, which is included as a potential predictor of stock returns. Campbell (1991), on the other hand, uses the model to analyse the unexpected component of returns, while Campbell and Shiller (1988a) concentrate on using the model to reinterpret the Marsh and Merton (1987) error-correction model of dividend behaviour in the context of a 'near-rational expectations' model in which dividends are persistent and prices are disturbed by persistent random noise.

Campbell and Shiller (1988b, 1988c) apply the dividend-ratio model to various data sets, including an updated Standard and Poor's. They find that the restrictions of the model tend to be rejected by the data and that the earnings variable is a powerful predictor of stock returns, particularly when returns are calculated over several years.

Example 7.6: The dividend–ratio model for UK equity prices

This model was applied to the UK data analysed in previous examples of this chapter. As a prerequisite, we require that ζ_t and ∇d_t are stationary: evidence that this is so was provided for ζ_t in example 3.1, while

Table 7.2 *Estimates of the dividend-ratio model for the UK*

Dependent variable	Explanatory variable		
	ζ_{t-1}	Δd_{t-1}	R^2
VAR(1) estimation			
ζ_t	0.963	0.093	0.927
	(0.015)	(0.250)	
Δd_t	−0.001	0.009	0.000
	(0.004)	(0.057)	
Returns regression			
$h_{1,t+1}$	0.082	−0.079	0.092
	(0.015)	(0.239)	

unit root tests confirm the stationarity of ∇d_t. A simple VAR(1) process fitted to the vector $\mathbf{z}_t$ yields the parameter estimates shown in table 7.2 and it is immediately clear that ζ_t does not Granger-cause ∇d_t.

Noting that $\mathbf{g}'=(1,0)$ and $\mathbf{h}'=(0,1)$, a Wald test of the restrictions (7.32) is numerically equivalent to a test that the coefficients in the regression of $h_{1,t+1}$ on ζ_{t-1} and ∇d_{t-1} are both zero, i.e., that returns are unforecastable. Computing $h_{1,t+1}$ using (7.29), with ρ estimated to be 0.953, yields the regression also shown in table 7.2. The coefficients are jointly significant, primarily because the lagged log dividend/price ratio has a significantly positive effect on returns: a finding that is consistent with that for previous US studies (see Campbell and Shiller, 1988c).

8 Future developments in the econometric modelling of financial time series

As the previous chapters have, hopefully, demonstrated, the econometric modelling of financial time series has proceeded apace in recent years. Such modelling has become increasingly popular for a number of interrelated reasons. Finance theory is perhaps unusual when compared with other fields of economics in that it often offers clear hypotheses to test using data that are both plentiful and accurately measured. Allied to this, financial data sets tend to have various features, for example non-normality, heterogeneity and non-linearity, that make them particularly popular with econometricians wishing to 'try out' empirically their latest models and estimation techniques. When one also takes into account the potential usefulness of financial models in directing investment and trading strategies, this popularity is thus hardly surprising.

We feel that it is appropriate to conclude our text with some personal views as to where the subject is likely to go next, not only within the academic fraternity of econometricians and finance theorists but also in the wider community of investment practitioners and financial market analysts.

Considering first the more technical developments, there is already evidence that some of the techniques discussed in previous chapters are being combined to produce, at least potentially, more powerful and flexible models. Two areas are of particular interest. The first is that of combining an unobserved component, or structural, model, with innovations that are allowed to follow ARCH processes: these are referred to as STARCH processes by Harvey, Ruiz and Sentana (1992). For example, the Muth (1960) model of chapter 3.2.1

$$x_t = z_t + u_t,$$

with

$$z_t = \mu + z_{t-1} + v_t$$

can be combined with ARCH(1) disturbances

$$u_t = U_{1t} h_{1t}^{1/2}$$

and

$$v_t = U_{2t} h_{2t}^{1/2},$$

where U_{1t} and U_{2t} are mutually independent $NID(0,1)$ processes and

$$h_{1t} = \alpha_{10} + \alpha_{11} u_{t-1}^2$$

and

$$h_{2t} = \alpha_{20} + \alpha_{21} v_{t-1}^2.$$

Various extensions of this simple STARCH model are possible. Chou, Engle and Kane (1992) consider a GARCH-M model in which the coefficient of the conditional variance, the parameter δ in equation (5.6), for example, changes over time according to a driftless random walk. Diebold and Nerlove (1989), on the other hand, analyse a latent factor model with ARCH effects in the factors. Such models seem to be particularly useful for analysing exchange rates.

The second general area is that of linking cointegration with other concepts in time series analysis. A natural extension of the cointegration framework set out in chapter 6 is to include deterministic components in the cointegrating regression. This allows the linear combinations of the variables which eliminate the unit roots to have non-zero linear trends, thus leading to the concept of *stochastic integration* introduced by Park (1992): see also Campbell and Perron (1991) and Hansen (1992). An even stronger definition of cointegration, called *deterministic cointegration* by Park, requires the same vectors which eliminate the unit roots to also eliminate the deterministic trends from the data. In fact, Campbell and Perron (1991) offer a more general definition of cointegration that, as well as incorporating both the above concepts, also allows some or all of the series to be trend stationary, unlike Engle and Granger's (1987) original definition which required all series to be difference stationary. Further extensions that have been proposed recently are the ideas of *regular* and *singular* stochastic cointegration (Park, 1992) and the concept of *multicointegration* (Granger and Lee, 1990): whether these extensions find a fruitful place in the modelling of financial time series is still a matter of some conjecture.

Granger and Hallman (1991) consider a non-linear generalisation of cointegration that uses the concept of a bivariate attractor and they propose some methods of estimation and testing. While this work is obviously in its infancy, such ideas could well have potential applications in modelling financial time series, particularly given the importance of non-linear models in this area, as is clearly shown in chapter 4.

On a somewhat related theme, the methodological controversy between classical and Bayesian techniques of testing for unit roots, briefly referred to in chapter 4, will certainly impinge on empirical work in finance. Indeed, it is already happening, with DeJong and Whiteman (1991a) presenting a Bayesian analysis of whether real stock prices and dividends are difference or trend stationary, an important prior assumption in the volatility debate of chapter 7, and Schotman and Van Dijk (1991) examining real exchange rates from a similar perspective. Since Bayesian techniques seem to present less conclusive evidence in favour of unit roots, this is a methodological debate of some substance for financial time series modellers. Bayesian approaches have also been proposed for inference within the ARCH class of models: see Geweke (1988, 1989).

Finally, alternatives to maximum likelihood estimation, such as generalised method of moments (Hansen, 1982, and Hansen and Singleton, 1982) and nonparametric methods (Gallant and Nychka, 1987) are becoming increasingly popular, particularly as such techniques are now being incorporated into econometric packages: see, for example, *MICROFIT 3.0*.

What about the impact such modelling techniques may have on the wider community of financial economists and market professionals? The non-linearity of many financial time series has opened the way for a reexamination of the methods of technical analysts, or chartists as they are often called. Neftci (1991) has initiated this line of research, demonstrating that only if a series is non-linear will technical trading rules be even potentially useful, but he also shows that rules based on comparisons of moving averages might capture some information that is ignored by standard (linear) prediction theory.

Many of the techniques discussed in this book are also becoming increasingly familiar with investment professionals. For example, a report by Geoghegan *et al.* (1992) to the UK Institute of Actuaries, concerning the actuarial use of a stochastic investment model developed by Wilkie (1986, 1987), discussed the concepts of VARs, ARCH and non-stationarity, before concluding

> that practitioners (both actuaries and non-actuaries) without the appropriate knowledge or understanding of econometric and time series models might enthrone a single particular model, and scenarios generated by it, with a degree of spurious accuracy... The proper application of this research cannot take place until a wider number of actuaries obtains the appropriate level of knowledge either through the examination system or by other suitable means. (Geoghegan *et al.*, 1992, p. 14)

Perhaps of most general importance is the increasingly common finding of predictability within financial markets: see Mills (1992b) for an informal review of recent evidence. For example, Gilles and LeRoy (1991) conclude from their survey of the excess volatility debate that asset-price volatility exceeds that predicted by the simplest of present-value models. Since, as Cochrane (1991b), for example, demonstrates, volatility tests are equivalent to long-horizon return regressions in which returns are forecasted by price/dividend ratios, excess volatility is equivalent to return forecastability, as found by authors such as Fama and French (1988) and Poterba and Summers (1988). Such forecastability, when placed alongside similar evidence gained from a variety of studies using a wide range of techniques (see Granger, 1993, for a survey of recent research), should be of obvious interest and importance to the financial community, for it again offers the possibility of developing profitable investment strategies and trading rules, something that the efficient markets hypothesis seemed to preclude. Of course, any strategy needs to be corrected for risk levels and transaction costs, and until such a strategy has been demonstrated, in an out-of-sample evaluation and after allowing for such corrections, to consistently produce positive profits, will there be worthwhile evidence against the hypothesis.

Nevertheless, the increasing importance of econometric and time series techniques in analysing financial data is abundantly clear, both in the relatively narrow academic field of financial economics and in the wider arena of financial markets themselves. It will surely remain an exciting area for economists to research in, even though the pace of development will surely render the present book out of date rather quickly!

Data appendix

The disk accompanying this book contains the following series:

LONG: Yield on 20 Year UK Gilts, quarterly, 1952Q1 to 1988Q4 (148 observations).

SHORT: 91 day UK Treasury Bill Rate, quarterly, 1952Q1 to 1988Q4 (148 observations).

FTAPRICE: *FTA All Share* price index, monthly, January 1965 to December 1990 (312 observations).

FTADIV: *FTA All Share* dividend index, monthly, January 1965 to December 1990 (312 observations).

FTARET: *FTA All Share* nominal returns, monthly, February 1965 to December 1990 (311 observations).

RPI: Retail Price Index, monthly, January 1965 to December 1990 (312 observations).

EXCH: Dollar/sterling exchange rate, weekly, January 1980 to December 1988 (470 observations).

BONDUS: U.S. bond yield, daily, 1 April 1986 to 29 December 1989 (960 observations).

BONDUK: U.K. bond yield, daily, 1 April 1986 to 29 December 1989 (960 observations).

BONDWG: West German bond yield, daily, 1 April 1986 to 29 December 1989 (960 observations).

BONDJP: Japanese bond yield, daily, 1 April 1986 to 29 December 1989 (960 observations).

References

Akaike, H. (1974), 'A New Look at the Statistical Model Identification', *IEEE Transactions on Automatic Control*, AC-19, 716–23.

Ali, M.M. and Giaccotto, C. (1982), 'The Identical Distribution Hypothesis for Stock Market Prices – Location and Scale-Shift Alternatives', *Journal of the American Statistical Association*, 77, 19–28.

Anderson, T.W. (1971), *The Statistical Analysis of Time Series*, New York: Wiley.

Ashley, R.A. and Patterson, D.M. (1986), 'A Nonparametric, Distribution-Free Test for Serial Independence in Stock Returns', *Journal of Financial and Quantitative Analysis*, 21, 221–7.

(1989), 'Linear versus Nonlinear Macroeconomies: a Statistical Test', *International Economic Review*, 30, 685–704.

Ashley, R.A., Patterson, D.M., and Hinich, M.J. (1986), 'A Diagnostic Test for Nonlinear Serial Dependence in Time Series Fitting Errors', *Journal of Time Series Analysis*, 7, 165–78.

Bachelier, L. (1900), 'Théorie de la Spéculation', *Annales de l'Ecole Normale Superieure*, Series 3, 17, 21–86.

Baillie, R.T. and Bollerslev, T. (1989), 'The Message in Daily Exchange Rates: a Conditional Variance Tale', *Journal of Business and Economic Statistics*, 7, 297–305.

(1992), 'Prediction in Dynamic Models with Time-Dependent Conditional Variances', *Journal of Econometrics*, 52, 91–113.

Baillie, R.T. and DeGennaro, R.P. (1990), 'Stock Returns and Volatility', *Journal of Financial and Quantitative Analysis*, 25, 203–14.

Baillie, R.T. and Myers, R.J. (1991), 'Bivariate GARCH Estimation of the Optimal Commodity Futures Hedge', *Journal of Applied Econometrics*, 6, 109–24.

Balke, N.S. and Fomby, T.B. (1991a), 'Shifting Trends, Segmented Trends, and Infrequent Permanent Shocks', *Journal of Monetary Economics*, 28, 61–85.

(1991b), 'Infrequent Permanent Shocks and the Finite-Sample Performance of Unit Root Tests', *Economics Letters*, 36, 269–73.

Berndt, E.R. (1991), *The Practice of Econometrics: Classic and Contemporary*, Reading, MA: Addison-Wesley.

Berndt, E.R., Hall, B.H., Hall, R.E. and Hausman, J.A. (1974), 'Estimation and

Inference in Nonlinear Structural Models', *Annals of Economic and Social Measurement*, 4, 653–65.

Beveridge, S. and Nelson, C.R. (1981), 'A New Approach to Decomposition of Economic Time Series into Permanent and Transitory Components with Particular Attention to Measurement of the "Business Cycle"', *Journal of Monetary Economics*, 7, 151–74.

Bhargava, A. (1986), 'On the Theory of Testing for Unit Roots in Observed Time Series', *Review of Economic Studies*, 53, 369–84.

Blanchard, O.J. and Watson, M.W. (1982), 'Bubbles, Rational Expectations, and Financial Markets', in P. Wachtel (ed.), *Crises in the Economic and Financial Structure*, Lexington, MA: Lexington Books, 295–315.

Bollerslev, T. (1986), 'Generalised Autoregressive Conditional Heteroskedasticity', *Journal of Econometrics*, 31, 307–27.

(1987), 'A Conditionally Heteroskedastic Time Series Model for Speculative Prices and Rates of Return', *Review of Economics and Statistics*, 69, 542–6.

(1988), 'On the Correlation Structure for the Generalized Autoregressive Conditional Heteroskedastic Process', *Journal of Time Series Analysis*, 9, 121–32.

Bollerslev, T. Chou, R.Y. and Kroner, K.F. (1992), 'ARCH Modelling in Finance: a Review of the Theory and Empirical Evidence', *Journal of Econometrics*, 52, 55–9.

Bollerslev, T., Engle, R.F. and Wooldridge, J.M. (1988), 'A Capital Asset Pricing Model with Time-Varying Covariances', *Journal of Political Economy*, 96, 116–31.

Box, G.E.P and Cox, D.R. (1964), 'An Analysis of Transformations', *Journal of the Royal Statistical Society*, Series B, 26, 211–43.

Box, G.E.P and Jenkins, G.M. (1976), *Time Series Analysis: Forecasting and Control*, Revised Edition, San Francisco: Holden Day.

Box, G.E.P. and Pierce, D.A. (1970), 'Distribution of Residual Autocorrelations in Autoregressive Moving Average Time Series Models', *Journal of the American Statistical Association*, 65, 1509–26.

Breedon, D.T. (1979), 'An Intertemporal Asset Pricing Model with Stochastic Consumption and Investment Opportunities', *Journal of Financial Economics*, 7, 265–96.

Brock, W.A. (1986), 'Distinguishing Random and Deterministic Systems: Abridged Version', *Journal of Economic Theory*, 40, 168–95.

(1988), 'Nonlinearity and Complex Dynamics in Economics and Finance', in P. Anderson, K. Arrow and D. Pines (eds.), *The Economy as an Evolving Complex System*, Reading, MA: SFI Studies in the Sciences of Complexity, 77–97.

Brock, W.A. and Baek, E.G. (1991), 'Some Theory of Statistical Inference for Nonlinear Science', *Review of Economic Studies*, 58, 697–716.

Brock, W.A. and Dechert, W.D. (1988), 'Theorems on Distinguishing Deterministic from Random Systems', in W.A. Barnett, E.R. Berndt and H. White (eds.), *Dynamic Econometric Modelling*, Cambridge University Press, 247–68.

Brock, W.A. and Sayers, C.L. (1988), 'Is the Business Cycle Characterized by Deterministic Chaos?', *Journal of Monetary Economics*, 22, 71–90.

Brockett, P.L., Hinich, M.J. and Patterson, D.M. (1988), 'Bispectral Based Tests for the Detection of Gaussianity and Linearity in Time Series', *Journal of the American Statistical Association*, 83, 657–64.

Brockwell, P.J. and Davis, R.A. (1991), *Time Series: Theory and Methods*, Second Edition, New York: Springer-Verlag.

Campbell, J.Y. (1991), 'A Variance Decomposition for Stock Returns', *Economic Journal*, 101, 157–79.

Campbell, J.Y. and Mankiw, N.G. (1987), 'Permanent and Transitory Components in Macroeconomic Fluctuations', *American Economic Review, Papers and Proceedings*, 77, 111–17.

Campbell, J.Y. and Perron, P. (1991), 'Pitfalls and Opportunities: What Macroeconomists Should Know About Unit Roots', Princeton University, Econometric Research Program, Research Memorandum No. 360.

Campbell, J.Y. and Shiller, R.J. (1987), 'Cointegration and Tests of Present Value Models', *Journal of Political Economy*, 95, 1062–88.

(1988a), 'Interpreting Cointegrated Models', *Journal of Economic Dynamics and Control*, 12, 503–22.

(1988b), 'Stock Prices, Earnings, and Expected Dividends', *Journal of Finance*, 43, 661–76.

(1988c), 'The Dividend–Price Ratio and Expectations of Future Dividends and Discount Factors', *Review of Financial Studies*, 1, 195–228.

Chesney, M. and Scott, L.O. (1989), 'Pricing European Currency Options: a Comparison of the Modified Black-Scholes Model and a Random Variance Model', *Journal of Financial and Quantitative Analysis*, 24, 267–84.

Chou, R.Y., Engle, R.F. and Kane, A. (1992), 'On the Measurement of Risk Aversion with Time-Varying Volatility and Unobservable Component of Wealth', *Journal of Econometrics*, 52, 201–24.

Chow, G.C. (1960), 'Tests of Equality Between Sets of Coefficients in Two Linear Regressions', *Econometrica*, 28, 591–605.

Christiano, L.J. (1988), 'Searching for Breaks in GNP', NBER Working Paper No. 2695.

Clark, P.K. (1973), 'A Subordinated Stochastic Process Model with Finite Variance for Speculative Prices', *Econometrica*, 41, 135–55.

Cochrane, J.H. (1988), 'How Big is the Random Walk in GNP?', *Journal of Political Economy*, 96, 893–920.

(1991a), 'A Critique of the Application of Unit Root Tests', *Journal of Economic Dynamics and Control*, 15, 275–84.

(1991b), 'Volatility Tests and Efficient Markets', *Journal of Monetary Economics*, 27, 463–85.

Cochrane. J.H. and Sbordone, A.M. (1988), 'Multivariate Estimates of the Permanent Components of GNP and Stock Prices', *Journal of Economic Dynamics and Control*, 12, 255–96.

Cooley, T.F. and LeRoy, S.F. (1985), 'Atheoretical Macroeconometrics: a Critique', *Journal of Monetary Economics*, 16, 283–308.

Cootner, P.A. (ed.) (1964), *The Random Character of Stock Market Prices*, Cambridge, MA: MIT Press.

Cowles, A. (1933), 'Can Stock Market Forecasters Forecast?', *Econometrica*, 1, 309–24.

(1944), 'Stock Market Forecasting', *Econometrica*, 12, 206–14.

(1960), 'A Revision of Previous Conclusions Regarding Stock Price Behaviour', *Econometrica*, 28, 909–15.

Cowles, A. and Jones, H.E. (1937), 'Some A Posteriori Probabilities in Stock Market Action', *Econometrica*, 5, 280–94.

Cramer, H. (1961), 'On Some Classes of Non-Stationary Processes', *Proceedings of the 4th Berkeley Symposium on Mathematical Statistics and Probability*, University of California Press, 57–78.

Davidson, J.E.H., Hendry, D.F., Srba, F. and Yeo, S. (1978), 'Econometric Modelling of the Aggregate Time Series Relationship between Consumers' Expenditure and Income in the United Kingdom', *Economic Journal*, 88, 661–92.

De Gooijer, J.G. (1989), 'Testing Non-linearities in World Stock Market Prices', *Economics Letters*, 31, 31–35.

DeJong, D.N., Nankervis, J.C., Savin, N.E. and Whiteman, C.H. (1992a), 'The Power Problems of Unit Root Tests in Time Series With Autoregressive Errors', *Journal of Econometrics*, 53, 323–43.

(1992b), 'Integration Versus Trend Stationarity in Time Series', *Econometrica*, 60, 423–33.

DeJong, D.N. and Whiteman, C.H. (1991a), 'The Temporal Stability of Dividends and Stock Prices: Evidence from the Likelihood Function', *American Economic Review*, 81, 600–17.

(1991b), 'Trends and Random Walks in Macroeconomic Time Series: A Reconsideration Based on the Likelihood Principle', *Journal of Monetary Economics*, 28, 221–54.

Diba, B.T. (1990), 'Bubbles and Stock-Price Volatility', in G.P. Dwyer and R.W. Hafer (eds.), *The Stock Market: Bubbles, Volatility, and Chaos*, Boston, MA: Kluwer Academic, 9–26.

Diba, B.T. and Grossman, H.I. (1987), 'On the Inception of Rational Bubbles', *Quarterly Journal of Economics*, 103, 697–700.

(1988), 'Explosive Rational Bubbles in Stock Prices?', *American Economic Review*, 78, 520–30.

Dickey, D.A., Bell, W.R. and Miller, R.B. (1986), 'Unit Roots in Time Series Models: Tests and Implications', *American Statistician*, 40, 12–26.

Dickey, D.A. and Fuller, W.A. (1979), 'Distribution of the Estimators for Autoregressive Time Series with a Unit Root', *Journal of the American Statistical Association*, 74, 427–31.

(1981), 'Likelihood Ratio Statistics for Autoregressive Time Series with a Unit Root', *Econometrica*, 49, 1057–72.

Dickey, D.A. and Pantula, S. (1987), 'Determining the Order of Differencing in Autoregressive Processes', *Journal of Business and Economic Statistics*, 5, 455–61.

Diebold, F.X. and Nerlove, M. (1989), 'The Dynamics of Exchange Rate Volatility: A Multivariate Latent ARCH Model', *Journal of Applied Econometrics*, 4, 1–21.

(1990), 'Unit Roots in Economic Time Series: a Selective Survey', in G.F. Rhodes and T.B. Fomby (eds.), *Advances in Econometrics, Volume 8*, Greenwich, CT: JAI Press, 3–69.

Diebold, F.X. and Rudebusch, G.D. (1989), 'Long Memory and Persistence in Aggregate Output', *Journal of Monetary Economics*, 24, 189–209.

Doan, T.A., Litterman, R.B. and Sims, C.A. (1984), 'Forecasting and Conditional Projection Using Realistic Prior Distributions', *Econometric Reviews*, 3, 1–100.

Dolado, J.J., Jenkinson, T. and Sosvilla-Rivero, S. (1990), 'Cointegration and Unit Roots', *Journal of Economic Surveys*, 4, 249–73.

Domowitz, I. and Hakkio, C.S. (1985), 'Conditional Variance and the Risk Premium in the Foreign Exchange Market', *Journal of International Economics*, 19, 47–66.

Engle, R.F. (1982), 'Autoregressive Conditional Heteroskedasticity with Estimates of the Variance of U.K. Inflation', *Econometrica*, 50, 987–1008.

Engle, R.F. and Bollerslev, T. (1986), 'Modelling the Persistence of Conditional Variances', *Econometric Reviews*, 5, 1–50.

Engle, R.F. and Granger, C.W.J. (1987), 'Cointegration and Error Correction: Representation, Estimation and Testing', *Econometrica*, 55, 251–76.

Engle, R.F., Hendry, D.F. and Richard, J.-F. (1983), 'Exogeneity', *Econometrica*, 51, 277–304.

Engle, R.F., Ito, T. and Lin, W.-L. (1990), 'Meteor Showers or Heat Waves? Heteroskedastic Intra Daily Volatility in the Foreign Exchange Market', *Econometrica*, 58, 525–42.

Engle, R.F., Lilien, D.M. and Robbins, R.P. (1987), 'Estimating Time Varying Risk Premia in the Term Structure: the ARCH-M Model', *Econometrica*, 55, 391–408.

Engle, R.F. and Yoo, B.S., (1987), 'Forecasting and Testing in Cointegrated Systems', *Journal of Econometrics*, 35, 143–59.

Epps, T.W. and Epps, M.L. (1976), 'The Stochastic Dependence of Security Price Changes and Transaction Volumes: Implications for the Mixture-of-Distribution Hypothesis', *Econometrica*, 44, 305–21.

Fama, E.F. (1965), 'The Behaviour of Stock-Market Prices', *Journal of Business*, 38, 34–105.

(1975), 'Short Term Interest Rates as Predictors of Inflation', *American Economic Review*, 65, 269–82.

Fama, E.F. and French, K.R. (1988), 'Dividend Yields and Expected Stock Returns', *Journal of Financial Economics*, 22, 3–25.

Fama, E.F. and MacBeth, J.D. (1973), 'Risk, Return, and Equilibrium: Empirical Tests', *Journal of Political Economy*, 81, 607–36.

Flavin, M.A. (1983), 'Excess Volatility in the Financial Markets: A Reassessment of the Empirical Evidence', *Journal of Political Economy*, 91, 929–56.

Frank, M. and Stengos, T. (1989), 'Measuring the Strangeness of Gold and Silver Rates of Return', *Review of Economic Studies*, 56, 553–67.

French, K.R., Schwert, G.W. and Stambaugh, R.F. (1987), 'Expected Stock Returns and Volatility', *Journal of Financial Economics*, 19, 3–29.

Fuller, W.A. (1976), *Introduction to Statistical Time Series*, New York: Wiley.

(1985), 'Nonstationary Autoregressive Time Series', in E.J. Hannan, P.R. Krishnaiah and M.M. Rao (eds.), *Handbook of Statistics, Volume 5: Time Series in the Time Domain*, Amsterdam: North-Holland, 1–24.

Gallant, A.R. and Nychka, D.W. (1987), 'Seminonparametric Maximum Likelihood Estimation', *Econometrica*, 55, 363–90.

Gallant, A.R. and White, H. (1988), *A Unified Theory of Estimation and Inference for Nonlinear Dynamic Models*, Oxford: Blackwell.

Geoghegan, T.J., Clarkson, R.S., Feldman, K.S., Green, S.J., Kitts, A., Lavecky, J.P., Ross, F.J.M., Smith, W.J. and Toutounchi, A. (1992), 'Report on the Wilkie Stochastic Investment Model', *Journal of the Institute of Actuaries*, 119, 1–40.

Geweke, J. (1978), 'Testing the Exogeneity Specification in the Complete Dynamic Simultaneous Equations Model', *Journal of Econometrics*, 7, 163–85.

(1982), 'Measurement of Linear Dependence and Feedback Between Time Series', *Journal of the American Statistical Association*, 79, 304–24.

(1984), 'Inference and Causality in Economic Time Series Models', in Z. Griliches and M.D. Intriligator (eds.), *Handbook of Econometrics, Volume II*, Amsterdam: North-Holland, 1101–44.

(1986), 'Modeling the Persistence of Conditional Variances: A Comment', *Econometric Reviews*, 5, 57–61.

(1988), 'Exact Inference in Models with Autoregressive Conditional Heteroskedasticity', in W.A. Barnett, E.R. Berndt and H. White (eds.), *Dynamic Econometric Modelling*, Cambridge University Press, 73–104.

(1989), 'Exact Predictive Densities in Linear Models with ARCH Disturbances', *Journal of Econometrics*, 44, 307–25.

Geweke, J. and Porter-Hudak, S. (1983), 'The Estimation and Application of Long Memory Time Series Models', *Journal of Time Series Analysis*, 4, 221–38.

Gibbons, M.R. (1982), 'Multivariate Tests of Financial Models', *Journal of Financial Economics*, 10, 3–27.

Gibbons, M.R., Ross, S.A. and Shanken, J. (1989), 'A Test of the Efficiency of a Given Portfolio', *Econometrica*, 57, 1121–52.

Gilles, C. and LeRoy, S.F. (1991), 'Econometric Aspects of the Variance-Bounds Tests: a Survey', *Review of Financial Studies*, 4, 753–91.

Godfrey, L.G. (1979), 'Testing the Adequacy of a Time Series Model', *Biometrika*, 66, 67–72.

(1988), *Misspecification Tests in Econometrics*, Cambridge University Press.

Granger, C.W.J. (1966), 'The Typical Spectral Shape of an Economic Variable', *Econometrica*, 34, 150–61.

(1969), 'Investigating Causal Relations by Econometric Models and Cross-Spectral Methods', *Econometrica*, 37, 424–38.

(1993), 'Forecasting Stock Market Prices: Lessons for Forecasters', *International Journal of Forecasting*, 9, 3–18.

Granger, C.W.J. and Andersen, A.P. (1978), *An Introduction to Bilinear Time Series Models*, Gottingen: Vandenhoeck and Ruprecht.

Granger, C.W.J. and Hallman, J.J. (1991), 'Long Memory Series with Attractors', *Oxford Bulletin of Economics and Statistics*, 53, 11–26.

Granger, C.W.J. and Joyeux, R. (1980), 'An Introduction to Long Memory Time Series Models and Fractional Differencing', *Journal of Time Series Analysis*, 1, 15–29.

Granger, C.W.J and Lee, T. (1990), 'Multicointegration', in G.F. Rhodes and T.B. Fomby (eds.), *Advances in Econometrics*, Vol. 8, Greenwich, CT: JAI Press, 71–84.

Granger, C.W.J. and Morgenstern, O. (1970), *Predictability of Stock Market Prices*, Heath: Lexington.

Granger, C.W.J. and Newbold, P. (1974), 'Spurious Regressions in Econometrics', *Journal of Econometrics*, 2, 111–120.

(1977), *Forecasting Economic Time Series*, New York: Academic Press.

Guilkey, D.K. and Schmidt, P. (1989), 'Extended Tabulations for Dickey-Fuller Tests', *Economics Letters*, 31, 355–7.

Haggan, V., Heravi, S.M. and Priestley, M.B. (1984), 'A Study of the Application of State-Dependent Models in Non-Linear Time Series Analysis', *Journal of Time Series Analysis*, 5, 69–102.

Haggan, V. and Ozaki, T. (1981), 'Modelling Non-Linear Vibrations Using an Amplitude-Dependent Autoregressive Time Series Model', *Biometrika*, 68, 189–96.

Hall, A. (1989), 'Testing for a Unit Root in the Presence of Moving Average Errors', *Biometrika*, 76, 49–56.

Hall, P. and Heyde, C.C. (1980), *Martingale Limit Theory and its Application*, New York: Academic Press.

Hall, S.G., Miles, D.K. and Taylor, M.P. (1989), 'Modelling Asset Prices with Time-Varying Betas', *Manchester School*, 57, 340–56.

Hamilton, J.D. (1986), 'On Testing for Self-Fulfilling Speculative Price Bubbles', *International Economic Review*, 27, 545–52.

(1989), 'A New Approach to the Economic Analysis of Nonstationary Time Series and the Business Cycle', *Econometrica*, 57, 357–84.

Hamilton, J.D. and Whiteman, C.H. (1985), 'The Observable Implications of Self-Fulfilling Expectations', *Journal of Monetary Economics*, 16, 353–73.

Hansen, B.E. (1992), 'Efficient Estimation and Testing of Cointegrating Vectors in the Presence of Deterministic Trends', *Journal of Econometrics*, 53, 87–121.

Hansen, L.P. (1982), 'Large Sample Properties of Generalized Method of Moments Estimators', *Econometrica*, 50, 1029–54.

Hansen, L.P. and Hodrick, R.J. (1980), 'Forward Exchange Rates as Optimal Predictors of Future Spot Rates', *Journal of Political Economy*, 88, 829–53.

Hansen, L.P. and Singleton, K.J. (1982), 'Generalized Instrumental Variables Estimation of Nonlinear Rational Expectations Models', *Econometrica*, 50, 1269–86.

Harvey, A.C. (1989), *Forecasting, Structural Time Series Models and the Kalman Filter*, Cambridge University Press.

Harvey, A.C., Ruiz, E. and Sentana, E. (1992), 'Unobserved Component Time Series Models with ARCH Disturbances', *Journal of Econometrics*, 52, 129–57.

Harvey, A.C. and Shephard, N. (1992), 'Structural Time Series Models', in G.S. Maddala, C.R. Rao and H.D. Vinod (eds.), *Handbook of Statistics, Volume 11:*

Econometrics, Amsterdam: North-Holland.
Hendry, D.F. (1987), 'Econometric Methodology: A Personal Perspective', in T.F. Bewley (ed.), *Advances in Econometrics – Fifth World Congress, Volume II*, Cambridge University Press, 29–48.
Hendry, D.F., Pagan, A.R. and Sargan, J.D. (1984), 'Dynamic Specification', in Z. Griliches and M.D. Intriligator (eds.), *Handbook of Econometrics, Volume II*, Amsterdam: North-Holland, 1023–100.
Higgins, M.L. and Bera, A.K. (1988), 'A Joint Test for ARCH and Bilinearity in the Regression Model', *Econometric Reviews*, 7, 171–81.
(1992), 'A Class of Nonlinear ARCH Models', *International Economic Review*, 33, 137–58.
Hinich, M.J. and Patterson, D.M. (1985), 'Evidence of Nonlinearity in Daily Stock Returns', *Journal of Business and Economic Statistics*, 3, 69–77.
(1989), 'Evidence of Nonlinearity in the Trade-by-Trade Stock Market Return Generating Process', in W.A. Barnett, J. Geweke and K. Shell (eds.), *Economic Complexity: Chaos, Sunspots, Bubbles, and Nonlinearity*, Cambridge University Press, 165–76.
Hosking, J.R.M. (1981), 'Fractional Differencing', *Biometrika*, 68, 165–76.
(1984), 'Modelling Persistence in Hydrological Time Series Using Fractional Differencing', *Water Resources Research*, 20, 1898–908.
Hsieh, D.A. (1988), 'The Statistical Properties of Daily Exchange Rates: 1974–1983', *Journal of International Economics*, 13, 171–86.
(1989a), 'Modeling Heteroskedasticity in Daily Foreign Exchange Rates', *Journal of Business and Economic Statistics*, 7, 307–17.
(1989b), 'Testing for Nonlinear Dependence in Daily Foreign Exchange Rates', *Journal of Business*, 62, 339–68.
(1991), 'Chaos and Nonlinear Dynamics: Application to Financial Markets', *Journal of Finance*, 46, 1839–77.
Hsu, D.A. (1977), 'Tests of Variance Shift at an Unknown Time Point', *Applied Statistics*, 26, 279–84.
(1979), 'Detecting Shifts of Parameter in Gamma Sequences with Applications to Stock Price and Air Traffic Flow Analysis', *Journal of the American Statistical Association*, 74, 31–40.
(1982), 'A Bayesian Robust Detection of Shift in the Risk Structure of Stock Market Returns', *Journal of the American Statistical Association*, 77, 29–39.
Huang, C.-F. and Litzenberger, R.H. (1988), *Foundations for Financial Economics*, Amsterdam: North-Holland.
Hurst, H. (1951), 'Long Term Storage Capacity of Reservoirs', *Transactions of the American Society of Civil Engineers*, 116, 770–99.
Hylleberg, S. and Mizon, G.E. (1989a), 'A Note on the Distribution of the Least Squares Estimation of a Random Walk with Drift', *Economics Letters*, 29, 225–30.
(1989b), 'Cointegration and Error Correction Mechanisms', *Economic Journal*, 99 (Supplement), 113–25.
Ingersoll, J.E. (1987), *Theory of Financial Decision Making*, Totowa, New Jersey: Rowman and Littlefield.

Jarque, C.M. and Bera, A.K. (1980), 'Efficient Tests for Normality, Homoskedasticity and Serial Dependence of Regression Residuals', *Economics Letters*, 6, 255–59.

Johansen, S. (1988), 'Statistical Analysis of Cointegration Vectors', *Journal of Economic Dynamics and Control*, 12, 231–54.

(1991), 'Estimation and Hypothesis Testing of Cointegrating Vectors in Gaussian Vector Autoregressive Models', *Econometrica*, 59, 1551–81.

Johansen, S. and Juselius, K. (1990), 'Maximum Likelihood Estimation and Inference on Cointegration – With Applications to the Demand for Money', *Oxford Bulletin of Economics and Statistics*, 52, 169–210.

Jorion, P. (1988), 'On Jump Processes in the Foreign Exchange and Stock Markets', *Review of Financial Studies*, 1, 427–45.

Judge, G.G., Griffiths, W.E., Carter Hill, R., Lutkepohl, H. and Lee, T.C. (1985), *The Theory and Practice of Econometrics*, Second Edition, New York: Wiley.

Keenan, D.M. (1985), 'A Tukey Nonadditivity-Type Test for Time Series Nonlinearity', *Biometrika*, 72, 39–44.

Kendall, M.J. (1953), 'The Analysis of Economic Time Series, Part I: Prices', *Journal of the Royal Statistical Society*, 96, 11–25.

Kleidon, A.W. (1986a), 'Variance Bounds Tests and Stock Price Valuation Models', *Journal of Political Economy*, 94, 953–1001.

(1986b), 'Bias in Small Sample Tests of Stock Price Rationality', *Journal of Business*, 59, 237–61.

Koop, G. (1992), '"Objective" Bayesian Unit Root Tests', *Journal of Applied Econometrics*, 7, 65–82.

Kunsch, H. (1986), 'Discrimination Between Monotonic Trends and Long-Range Dependence', *Journal of Applied Probability*, 23, 1025–30.

Lam, P.S. (1990), 'The Hamilton Model with a General Autoregressive Component: Estimation and Comparison with Other Models of Economic Time Series', *Journal of Monetary Economics*, 20, 409–32.

Leamer, E.E. (1985), 'Vector Autoregressions for Causal Inference?', *Carnegie-Rochester Conference Series on Public Policy*, 22, 255–304.

LeRoy, S.F. (1982), 'Expectations Models of Asset Prices: A Survey of Theory', *Journal of Finance*, 37, 185–217.

(1989), 'Efficient Capital Markets and Martingales', *Journal of Economic Literature*, 27, 1583–621.

LeRoy, S.F. and Porter, R.D. (1981), 'The Present Value Relation: Tests Based on Implied Variance Bounds', *Econometrica*, 49, 555–74.

Ljung, G.M. and Box, G.E.P. (1978), 'On a Measure of Lack of Fit in Time Series Models', *Biometrika*, 66, 67–72.

Lo, A.W. (1991), 'Long-Term Memory in Stock Market Prices', *Econometrica*, 59, 1279–313.

Lo, A.W. and MacKinlay, A.C. (1988), 'Stock Prices do not Follow Random Walks: Evidence from a Simple Specification Test', *Review of Financial Studies*, 1, 41–66.

(1989), 'The Size and Power of the Variance Ratio Test in Finite Samples: A Monte Carlo Investigation', *Journal of Econometrics*, 40, 203–38.

Lutkepohl, H. (1985), 'Comparison of Criteria for Estimating the Order of a Vector Autoregressive Process', *Journal of Time Series Analysis*, 6, 35–52.

(1991), *Introduction to Multiple Time Series Analysis*, Berlin: Springer-Verlag.

MacDonald, R. and Speight, A.E.H. (1988), 'The Term Structure of Interest Rates in the UK', *Bulletin of Economic Research*, 40, 287–99.

MacKinlay, A.C. (1987), 'On Multivariate Tests of the CAPM', *Journal of Financial Economics*, 18, 341–71.

MacKinnon, J.G. (1990), 'Critical Values for Cointegration Tests', UC San Diego Discussion Paper 90–4.

McKenzie, E. (1988), 'A Note on Using the Integrated Form of ARIMA Forecasts', *International Journal of Forecasting*, 4, 117–24.

McLeod, A.J. and Li, W.K. (1983), 'Diagnostic Checking ARMA Time Series Models Using Squared-Residual Correlations', *Journal of Time Series Analysis*, 4, 269–73.

Mandelbrot, B.B. (1963), 'New Methods in Statistical Economics', *Journal of Political Economy*, 71, 421–40.

(1966), 'Forecasts of Future Prices, Unbiased Markets and "Martingale" Models', *Journal of Business*, 39, 242–55.

(1969), 'Long-Run Linearity, Locally Gaussian Process, *H*-Spectra, and Infinite Variances', *International Economic Review*, 10, 82–111.

(1972), 'Statistical Methodology for Nonperiodic Cycles: From the Covariance to R/S Analysis', *Annals of Economic and Social Measurement*, 1/3, 259–90.

(1989), 'Louis Bachelier', in J. Eatwell, M. Milgate and P. Newman (eds.), *The New Palgrave: Finance*, London: Macmillan, 86–8.

Mandelbrot, B.B. and Wallis, J.R. (1969), 'Some Long-Run Properties of Geophysical Records', *Water Resources Research*, 5, 321–40.

Mankiw, N.G., Romer, D. and Shapiro, M.D. (1985), 'An Unbiased Reexamination of Stock Market Volatility', *Journal of Finance*, 40, 677–87.

Maravall, A. (1983), 'An Application of Nonlinear Time Series Forecasting', *Journal of Business and Economic Statistics*, 3, 350–5.

Marsh, T.A. and Merton, R.C. (1987), 'Dividend Behaviour for the Aggregate Stock Market', *Journal of Business*, 60, 1–40.

Mattey, J. and Meese, R.A. (1986), 'Empirical Assessment of Present Value Relations', *Econometric Reviews*, 5, 171–234.

Meese, R.A. (1986), 'Testing for Bubbles in Exchange Markets: A Case of Sparkling Rates?', *Journal of Political Economy*, 94, 345–73.

Meese, R.A. and Singleton, K.J. (1982), 'On Unit Roots and the Empirical Modeling of Exchange Rates', *Journal of Finance*, 37, 1029–35.

Melino, A. and Turnbull, S.M. (1990), 'Pricing Foreign Currency Options with Stochastic Volatility', *Journal of Econometrics*, 45, 239–65.

Menzefricke, U. (1981), 'A Bayesian Analysis of a Change in the Precision of a Sequence of Independent Random Normal Variables at an Unknown Time Point', *Applied Statistics*, 30, 141–6.

Merton, R.C. (1973), 'An Intertemporal Capital Asset Pricing Model', *Econometrica*, 41, 867–87.

(1980), 'On Estimating the Expected Return on the Market: An Exploratory

Investigation', *Journal of Financial Economics*, 8, 323–61.

Milhøj, A. (1985), 'The Moment Structure of ARCH Processes', *Scandinavian Journal of Statistics*, 12, 281–92.

(1987), 'A Conditional Variance Model for Daily Deviations of an Exchange Rate', *Journal of Business and Economic Statistics*, 5, 99–103.

Mills, T.C. (1990), *Time Series Techniques for Economists*, Cambridge University Press.

(1991a), 'Assessing the Predictability of U.K. Stock Market Returns Using Statistics Based on Multiperiod Returns', *Applied Financial Economics*, 1, 241–5.

(1991b), 'Equity Prices, Dividends and Gilt Yields in the UK: Cointegration, Error Correction and "Confidence"', *Scottish Journal of Political Economy*, 38, 242–55.

(1991c), 'The Term Structure of UK Interest Rates: Tests of the Expectations Hypothesis', *Applied Economics*, 23, 599–606.

(1992a), 'Testing the Present Value Model of Equity Prices for the UK Stock Market', *Journal of Business Finance and Accounting*, 20.

(1992b), 'Predicting the Unpredictable: Science and Guesswork in Financial Market Forecasting', Institute of Economic Affairs, Occasional Paper 87.

Mills, T.C. and Mills, A.G. (1991), 'The International Transmission of Bond Market Movements', *Bulletin of Economic Research*, 43, 273–82.

Mills, T.C. and Stephenson, M.J. (1985), 'An Empirical Analysis of the U.K. Treasury Bill Market', *Applied Economics*, 17, 689–703.

Morgan, A. and Morgan, I. (1987), 'Measurement of Abnormal Returns from Small Firms', *Journal of Business and Economic Statistics*, 5, 121–9.

Muth, J.F. (1960), 'Optimal Properties of Exponentially Weighted Forecasts', *Journal of the American Statistical Association*, 55, 299–305.

Neftci, S.N. (1991), 'Naive Trading Rules in Financial Markets and Weiner-Kolmogorov Prediction Theory: A Study of "Technical Analysis"', *Journal of Business*, 64, 549–71.

Nelson, C.R. (1988), 'Spurious Trend and Cycle in the State Space Decomposition of a Time Series with a Unit Root', *Journal of Economic Dynamics and Control*, 12, 476–88.

Nelson, C.R. and Plosser, C.I. (1982), 'Trends and Random Walks in Macroeconomic Time Series: Some Evidence and Implications', *Journal of Monetary Economics*, 10, 139–62.

Nelson, C.R. and Schwert, G.W. (1977), 'Short-Term Interest Rates as Predictors of Inflation: On Testing the Hypothesis that the Real Rate of Interest is Constant', *American Economic Review*, 67, 478–86.

Nelson, D.B. (1990), 'ARCH Models as Diffusion Approximations', *Journal of Econometrics*, 45, 7–38.

(1991), 'Conditional Heteroskedasticity in Asset Returns', *Econometrica*, 59, 347–70.

Nerlove, M., Grether, D.M. and Carvalho, J.L. (1979), *Analysis of Economic Time Series: A Synthesis*, New York: Academic Press.

Newey, W.K. and West, K.D. (1987), 'A Simple Positive Semidefinite, Heteroskedasticity Consistent Covariance Matrix', *Econometrica*, 55, 703–8.

Osborne, M.M. (1959), 'Brownian Motion in the Stock Market', *Operations Research*, 7, 145–73.

Pagan, A.R. and Schwert, G.W. (1990), 'Alternative Models for Conditional Stock Volatility', *Journal of Econometrics*, 45, 267–90.

Pagan, A.R. and Wickens, M.R. (1989), 'A Survey of Some Recent Econometric Methods', *Economic Journal*, 99, 962–1025.

Pantula, S.G. (1986), 'Modelling the Persistence of Conditional Variances: a Comment', *Econometric Reviews*, 5, 71–4.

Pantula, S.G. and Hall, A. (1991), 'Testing for Unit Roots in Autoregressive Moving Average Models: An Instrumental Variable Approach', *Journal of Econometrics*, 48, 325–53.

Park, J.Y. (1990), 'Testing for Unit Roots and Cointegration by Variable Addition', in G.F. Rhodes and T.B. Fomby (eds.), *Advances in Econometrics*, Volume 8, Greenwich, CT: JAI Press, 107–33.

(1992), 'Canonical Cointegrating Regressions', *Econometrica*, 60, 119–43.

Park, J.Y. and Phillips, P.C.B. (1988), 'Statistical Inference in Regressions with Cointegrated Processes: Part I', *Econometric Theory*, 4, 468–97.

Pearson, K. and Rayleigh, Lord (1905), 'The Problem of the Random Walk', *Nature*, 72, 294, 318, 342.

Perron, P. (1988), 'Trends and Random Walks in Macroeconomic Time Series: Further Evidence from a New Approach', *Journal of Economic Dynamics and Control*, 12, 297–332.

(1989a), 'The Great Crash, the Oil Price Shock, and the Unit Root Hypothesis', *Econometrica*, 57, 1361–401.

(1989b), 'Testing for a Unit Root in a Time Series with a Changing Mean', *Journal of Business and Economic Statistics*, 8, 153–62.

(1991), 'Test Consistency with Varying Sampling Frequency', *Econometric Theory*, 7, 341–68.

Pesaran, M.H. and Pesaran, B. (1991), *Microfit 3.0*, Oxford: Oxford University Press.

Phillips, P.C.B. (1986), 'Understanding Spurious Regressions in Econometrics', *Journal of Econometrics*, 33, 311–40.

(1987a), 'Time Series Regression with a Unit Root', *Econometrica*, 55, 227–301.

(1987b), 'Towards a Unified Asymptotic Theory for Autoregression', *Biometrika*, 74, 535–47.

(1987c), 'Asymptotic Expansions in Nonstationary Vector Autoregressions', *Econometric Theory*, 3, 45–68.

(1988), 'Reflections on Econometric Methodology', *Economic Record*, 64, 344–59.

(1991a), 'To Criticize the Critics: An Objective Bayesian Analysis of Stochastic Trends', *Journal of Applied Econometrics*, 6, 333–64.

(1991b), 'Bayesian Routes and Unit Roots: De Rebus Prioribus Semper est Disputandum', *Journal of Applied Econometrics*, 6, 435–74.

(1991c), 'Optimal Inference in Cointegrated Systems', *Econometrica*, 59, 283–306.

Phillips, P.C.B. and Durlauf, S.N. (1986), 'Multiple Time Series Regression with Integrated Processes', *Review of Economic Studies*, 53, 473–96.

Phillips, P.C.B. and Hansen, B.E. (1990), 'Statistical Inference in Instrumental Variables Regression with *I*(1) Processes', *Review of Economic Studies*, 57, 99–125.

Phillips, P.C.B. and Loretan, M. (1991), 'Estimating Long-Run Economic Equilibria', *Review of Economic Studies*, 58, 407–36.

Phillips, P.C.B. and Ouliaris, S. (1988), 'Testing for Cointegration Using Principal Components Methods', *Journal of Economic Dynamics and Control*, 12, 205–30.

(1990), 'Asymptotic Properties of Residual Based Tests for Cointegration', *Econometrica*, 58, 165–94.

Phillips, P.C.B. and Perron, P. (1988), 'Testing for Unit Roots in Time Series Regression', *Biometrika*, 75, 335–46.

Pierce, D.A. (1979), 'Signal Extraction Error in Nonstationary Time Series', *Annals of Statistics*, 7, 1303–20.

Plosser, C.I. and Schwert, G.W. (1978), 'Money, Income, and Sunspots: Measuring Economic Relationships and the Effects of Differencing', *Journal of Monetary Economics*, 4, 637–60.

Poskitt, D.S. and Tremayne, A.R. (1987), 'Determining a Portfolio of Linear Time Series Models', *Biometrika*, 74, 125–37.

Poterba, J.M. and Summers, L.H. (1988), 'Mean Reversion in Stock Prices: Evidence and Implications', *Journal of Financial Economics*, 22, 27–59.

Priestley, M.B. (1980), 'State-Dependent Models: A General Approach to Nonlinear Time Series Analysis', *Journal of Time Series Analysis*, 1, 47–71.

Ramsey, J.B. (1969), 'Tests for Specification Errors in Classical Linear Least Squares Regression Analysis', *Journal of the Royal Statistical Society*, Series B, 31, 350–71.

(1990), 'Economic and Financial Data as Nonlinear Processes', in G.P. Dwyer and R.W. Hafer (eds.), *The Stock Market: Bubbles, Volatility, and Chaos*, Boston, MA: Kluwer Academic, 81–134.

Ramsey, J.B., Sayers, C.L. and Rothman, P. (1990), 'The Statistical Properties of Dimension Calculations Using Small Data Sets: Some Economic Applications', *International Economic Review*, 31, 991–1020.

Ramsey, J.B. and Yuan, H.-J. (1990), 'The Statistical Properties of Dimension Calculations Using Small Data Sets', *Nonlinearity*, 3, 155–76.

Rappoport, P. and Reichlin, L. (1989), 'Segmented Trends and Non-Stationary Time Series', *Economic Journal*, 99 (Supplement), 168–77.

Reichlin, L. (1989), 'Structural Change and Unit Root Econometrics', *Economics Letters*, 31, 231–3.

Richardson, M. and Stock, J.H. (1989), 'Drawing Inferences from Statistics Based on Multi-Year Asset Returns', *Journal of Financial Economics*, 25, 323–48.

Roberts, H.V. (1959), 'Stock-Market "Patterns" and Financial Analysis: Methodological Suggestions', *Journal of Finance*, 14, 1–10.

Robinson, P. (1977), 'The Estimation of a Non-Linear Moving Average Model',

Stochastic Processes and Their Applications, 5, 81–90.

Said, S.E. and Dickey, D.A. (1984), 'Testing for Unit Roots in Autoregressive Moving-Average Models with Unknown Order', *Biometrika*, 71, 599–607.

(1985), 'Hypothesis Testing in ARIMA(p,1,q) Models', *Journal of the American Statistical Association*, 80, 369–74.

Saikkonen, P. (1991), 'Asymptotically Efficient Estimation of Cointegrating Regressions', *Econometric Theory*, 7, 1–21.

Samuelson, P.A. (1965), 'Proof That Properly Anticipated Prices Fluctuate Randomly', *Industrial Management Review*, 6, 41–9.

(1973), 'Proof That Properly Discounted Present Values of Assets Vibrate Randomly', *Bell Journal of Economics and Management Science*, 4, 369–74.

Sargan, J.D. and Bhargava, A.S. (1983), 'Testing Residuals from Least Squares Regression for being Generated by the Gaussian Random Walk', *Econometrica*, 51, 153–74.

Scheinkman, J.A. (1990), 'Nonlinearities in Economic Dynamics', *Economic Journal*, 100 (Supplement), 33–48.

Scheinkman, J.A. and LeBaron, B. (1989), 'Nonlinear Dynamics and Stock Returns', *Journal of Business*, 62, 311–37.

Schmidt, P. (1990), 'Dickey-Fuller Tests with Drift', in G.F. Rhodes and T.B. Fomby (eds.), *Advances in Econometrics, Volume 8*, Greenwich, CT.: JAI Press, 161–200.

Schotman, P. and Van Dijk, H.K. (1991), 'A Bayesian Analysis of the Unit Root in Real Exchange Rates', *Journal of Econometrics*, 49, 195–238.

Schwarz, G. (1978), 'Estimating the Dimension of a Model', *Annals of Statistics*, 6, 461–4.

Schwert, G.W. (1987), 'Effects of Model Specification on Tests for Unit Roots in Macroeconomic Data', *Journal of Monetary Economics*, 20, 73–105.

Scott, L.O. (1987), 'Option Pricing when the Variance Changes Randomly. Theory, Estimation and an Application', *Journal of Financial and Quantitative Analysis*, 22, 419–38.

Shiller, R.J. (1979), 'The Volatility of Long Term Interest Rates and Expectations Models of the Term Structure', *Journal of Political Economy*, 87, 1190–209.

(1981a), 'Do Stock Prices Move Too Much to be Justified by Subsequent Changes in Dividends?', *American Economic Review*, 71, 421–36.

(1981b), 'The Use of Volatility Measures in Assessing Market Efficiency', *Journal of Finance*, 36, 291–304.

Shiller, R.J. and Perron, P. (1985), 'Testing the Random Walk Hypothesis: Power Versus Frequency of Observation', *Economics Letters*, 18, 381–6.

Sims, C.A. (1980), 'Macroeconomics and Reality', *Econometrica*, 48, 1–48.

(1981), 'An Autoregressive Index Model for the US 1948–1975', in J. Kmenta and J.B. Ramsey (eds.), *Large-Scale Macroeconometric Models*, Amsterdam: North-Holland, 283–327.

(1982), 'Policy Analysis with Econometric Models', *Brookings Papers on Economic Activity*, 1, 107–64.

(1987), 'Making Economics Credible', in T.F. Bewley (ed.), *Advances in Econo-*

metrics – Fifth World Congress, Volume II, Cambridge University Press, 46–60.
(1988), 'Bayesian Skepticism of Unit Root Econometrics', *Journal of Economic Dynamics and Control*, 12, 463–75.

Sims, C.A. and Uhlig, H. (1991), 'Understanding Unit Rooters: A Helicopter Tour', *Econometrica*, 59, 1591–9.

Sowell, F. (1992a), 'Maximum Likelihood Estimation of Stationary Univariate Fractionally Integrated Time Series Models', *Journal of Econometrics*, 53, 165–88.

(1992b), 'Modeling Long-run Behaviour with the Fractional ARIMA Model', *Journal of Monetary Economics*, 29, 277–302.

Spanos, A. (1986), *Statistical Foundations of Econometric Modelling*, Cambridge University Press.

Stock, J.H. (1987), 'Asymptotic Properties of Least Squares Estimators of Cointegrating Vectors', *Econometrica*, 55, 1035–56.

(1991), 'Confidence Intervals for the Largest Autoregressive Root in U.S Macroeconomic Time Series', *Journal of Monetary Economics*, 28, 435–59.

Stock, J.H. and Watson, M.W. (1988a), 'Variable Trends in Economic Time Series', *Journal of Economic Perspectives*, 2, 147–74.

(1988b), 'Testing for Common Trends', *Journal of the American Statistical Association*, 83, 1097–107.

Subba Rao, T. (1981), 'On the Theory of Bilinear Models', *Journal of the Royal Statistical Society*, Series B, 43, 244–55.

Subba Rao, T. and Gabr, M.M. (1984), *An Introduction to Bispectral Analysis and Bilinear Time Series Models*, Berlin: Springer-Verlag.

Taylor, S. (1986), *Modelling Financial Time Series*, New York: Wiley.

Tauchen, G.E. and Pitts, M. (1983), 'The Price Variability–Volume Relationship on Speculative Markets', *Econometrica*, 51, 485–505.

Thaler, R. (1987a), 'The January Effect', *Journal of Economic Perspectives*, 1(1), 197–201.

(1987b), 'Seasonal Movements in Security Prices II: Weekend, Holiday, Turn of the Month, and Intraday Effects', *Journal of Economic Perspectives*, 1(2), 169–77.

Tirole, J. (1985), 'Asset Bubbles and Overlapping Generations', *Econometrica*, 53, 1071–100.

Tong, H. and Lim, K.S. (1980), 'Threshold Autoregression, Limit Cycles, and Cyclical Data', *Journal of the Royal Statistical Society*, Series B, 42, 245–92.

Tsay, R.S. (1986), 'Nonlinearity Tests for Time Series', *Biometrika*, 73, 461–6.

Wecker, W.E. (1981), 'Asymmetric Time Series', *Journal of the American Statistical Association*, 76, 16–21.

Weiss, A.A. (1984), 'ARMA Models with ARCH Errors', *Journal of Time Series Analysis*, 5, 129–43.

(1986a), 'Asymptotic Theory for ARCH Models: Estimation and Testing', *Econometric Theory*, 2, 107–31.

(1986b), 'ARCH and Bilinear Time Series Models: Comparison and Combination', *Journal of Business and Economic Statistics*, 4, 59–70.

West, K.D. (1987), 'A Specification Test for Speculative Bubbles', *Quarterly Journal of Economics*, 102, 553–80.

(1988a), 'Asymptotic Normality, When Regressors have a Unit Root', *Econometrica*, 56, 1397–418.

(1988b), 'Dividend Innovations and Stock Price Volatility', *Econometrica*, 56, 37–61.

(1988c), 'Bubbles, Fads and Stock Price Volatility Tests: A Partial Evaluation', *Journal of Finance*, 43, 636–56.

White, H. (1980), 'A Heteroskedasticity-Consistent Covariance Matrix Estimator and a Direct Test for Heteroskedasticity', *Econometrica*, 48, 817–38.

(1984), *Asymptotic Theory for Econometricians*, New York: Academic Press.

White, H. and Domowitz, I. (1984), 'Nonlinear Regression with Dependent Observations', *Econometrica*, 52, 643–61.

Wiggins, J.B. (1987), 'Option Values Under Stochastic Volatility. Theory and Empirical Estimates', *Journal of Financial Economics*, 19, 351–72.

Wilkie, A.D. (1986), 'A Stochastic Investment Model for Actuarial Use', *Transactions of the Faculty of Actuaries*, 39, 341.

(1987), 'Stochastic Investment Models – Theory and Application', *Insurance: Mathematics and Economics*, 6, 65.

Wold, H. (1938), *A Study in the Analysis of Stationary Time Series*, Stockholm: Almqvist and Wiksell.

Working, H. (1934), 'A Random-Difference Series for Use in the Analysis of Time Series', *Journal of the American Statistical Association*, 29, 11–24.

(1960), 'Note on the Correlation of First Differences of Averages in a Random Chain', *Econometrica*, 28, 916–18.

Zivot, E. and Andrews, D.W.K. (1990), 'Further Evidence on the Great Crash, the Oil Price Shock, and the Unit Root Hypothesis', Cowles Foundation Discussion Paper No. 944.

Index

A